International Journal on Advances in Life Sciences

2017 vol. 9 nr. 1&2

Printed from e-media with permission by:

Curran Associates, Inc.
57 Morehouse Lane
Red Hook, NY 12571

Some format issues inherent in the e-media version may also appear in this print version.

Printed by Curran Associates, Inc. (2018)

International Academy, Research, and Industry Association (IARIA)
412 Derby Way
Wilmington, DE 19810

Phone: (408) 893-6407
Fax: (408) 527-6351

petre@iaria.org

Additional copies of this publication are available from:

Curran Associates, Inc.
57 Morehouse Lane
Red Hook, NY 12571 USA
Phone: 845-758-0400
Fax: 845-758-2633
Email: curran@proceedings.com
Web: www.proceedings.com

The *International Journal on Advances in Life Sciences* is published by IARIA.
ISSN: 1942-2660
journals site: http://www.iariajournals.org
contact: petre@iaria.org

Sponsored by IARIA
www.iaria.org

International Journal on Advances in Life Sciences
Volume 9, Number 1 & 2, 2017

Stephen White, University of Huddersfield, UK
Levent Yilmaz, Auburn University, USA
Eiko Yoneki, University of Cambridge, UK

International Journal on Advances in Life Sciences
Volume 9, Numbers 1 & 2, 2017

CONTENTS

Assessing an Electronic Health Record (EHR): How Do Basic Assumptions in Traditional Health Technology Assessment (HTA), and Empirical Features Fit?

Anne Granstrøm Ekeland
University Hospital of North Norway
Norwegian Centre for E-health Research
Tromsø, Norway
E-mail: anne.granstrom.ekeland@ehealthresearch.no

Gunnar Ellingsen
Telemedicine and e-health Research Group
UIT – The Arctic University of Norway
Tromsø, Norway
E-mail: gunnar.ellingsen@uit.no

***Abstract*—The use of health technology assessment (HTA) in the field of information and communication technologies (ICT) is receiving increasing attention. However, assessments this far are limited. In HTA, randomized controlled trials (RCTs) are the gold standard approach, building on a coherent set of basic philosophical assumptions. Scholars have raised questions concerning the assumptions and their fit with e-health, and thus questioned the ability of HTAs to produce useful assessments. The failure of assumptions to reflect empirical features of e-health is one explanation. This paper discusses this tension. Based upon the conference paper "Assessing Electronic Health Records: Are Basic Assumptions in Health Technology Assessment Useful?," presented at The Eight International Conference on eHealth, Telemedicine, and Social Medicine eTELEMED 2016, April 24–28, 2016, in Venice, Italy, we have elaborated the empirical substantiation and present an extended version. Using a sociotechnical perspective, we studied a large-scale electronic health record program in northern Norway. Drawing on data over a 5-year period, we discuss how the program's plans, organization, and activities correspond to RCT assumptions. We found that the RCT assumptions of a stable world, fixed interventions, and controlled implementation processes differ substantially from the real-life processes. Thus, RCT approaches that build on such assumptions fail to address important features of the program and fail to produce knowledge that fully demonstrates (the causes of) empirical benefits or pitfalls. As a result, we suggest embedding a world in flux in the assumptions of HTA where social, technical, and clinical entities continuously shape each other in dynamic processes. This may increase the relevance of HTA in ICT implementation projects.**

Keywords-health technology assessment (HTA); approaches and methods in randomized controlled trials (RCT); empirical features of electronic health records; assumptions in constructive assessments.

I. INTRODUCTION

A. *Background*

This paper is an extended version of a paper presented at the Eighth International Conference on eHealth, Telemedicine, and Social Medicine, eTELEMED 2016, April 24–28, 2016, in Venice, Italy [1]. The six-page conference paper allowed limited space for empirical documentation and data from participation in an implementation process of the electronic health record program (FIKS). In this extended paper, the empirical documentation is presented to substantiate the argument in the conference paper, for novel and more relevant assumptions to guide assessment approaches within the health technology assessment (HTA) tradition. In addition, the account in the conference paper of the steps within the HTA communities was very broad. Therefore, in this extended paper, we also describe and discuss in detail the policy and scientific steps within segments of this community. The paper provides an account of processes initiated by the International Health Technology Assessment network (HTAi) to meet challenges concerning assessment.

By expanding the authorship and time period of the data material to include the year of 2011, the extended paper also incorporates novel detailed empirical content to strengthen the argument of the conference paper. This paper also refers to previously published results that provide new perspectives on the empirical development of information and communication technology (ICT) in health care.

The new material is in line with the original topic but substantiates and expands the argument. The conclusion is updated to capture the new content by suggesting steps for updating the theoretical assumptions behind HTA approaches.

This paper is, similarly to the original, positioned within a sociotechnical perspective and seeks to produce insights into the compatibility between the assumptions behind HTA approaches to assessments and the practices that are assessed. The main idea is that there has to be coherence to produce relevant knowledge.

The need for assessments of ICT programs has been strongly expressed. For instance, the Parliament in the United Kingdom (UK) stated in a summary of the National Health Service (NHS) information technology (IT) program: "The original objective was to ensure every NHS patient had an individual electronic care record which could be rapidly transmitted between different parts of the NHS, in order to make accurate patient records available to NHS staff at all times. This intention has proved beyond the capacity of the department to deliver and the department is no longer delivering a universal system. Implementation of alternative up-to-date IT systems has fallen significantly behind schedule and costs have escalated" [2].

Health technology assessments are designed for, and expected to, produce knowledge to help decide about and procure technology and services that are accurate and cost-effective and have the expected value and quality [3]. The Norwegian health authorities and international scientific networks for conducting HTAs have called for steps to strengthen its use in ICT. In 2016, the Regional Health Authority, North Norway (NNHA) funded a three-year project for developing and adapting HTA approaches and tools: HTA for ICT [4]. The project builds on the "One Patient–One Record" white paper presented to the Norwegian Parliament [5]. This extended paper is part of the HTA for the ICT project.

The need to adapt and develop assessments for e-health has also been expressed in several scientific publications; some are referred to in Section C. A common concern is that established assessments have weaknesses in that they produce less relevant and timely knowledge. In this paper, weaknesses connected to basic theoretical and philosophical assumptions in HTA are addressed, more specifically, those expressed in the gold standard approach of randomized controlled trials (RCTs) and related to the development of electronic health records (EHRs) in northern Norway.

The research question is how assumptions of RCT are amenable to empirical features of the Common Implementation of Clinical Systems (the Norwegian acronym is FIKS), a large-scale program for developing and implementing a new EHR [6][7]. The paper also briefly comments on RCT approaches and methods, because they rely on the same set of basic assumptions. FIKS started in 2012 and was scheduled to last through 2016. It lies within the jurisdiction of the NNHA in North Norway. The goal is to establish a common electronic patient record for all hospitals in the northern region of Norway. No pre-implementation or baseline evaluation was carried out.

The first objective of the investigation is to contribute to a knowledge base for dealing with challenges experienced when conducting traditional HTAs for ICT. The second objective is to briefly present and substantiate alternative assumptions and assessment designs. The alternative assumptions are presented as logical consequences of the study's empirical findings. They are discussed as capable of strengthening HTA use in ICT for the benefit of patients, health professionals, policy makers, leaders, and industry.

In Sections B and C, HTAs, represented by RCT assumptions and weaknesses, are presented. In Section II, an account of FIKS is given, followed by the methods and materials. The results and discussion in Section III are divided into three sub-sections; each addresses different assumptions in RCT: a singular reality (context), a clear definition of the intervention, and a controlled implementation process. Approaches and methods that accommodate different assumptions (a reality in flux and interventions and implementation as ongoing socio-, technical, and medical achievements) are discussed in Section IV. In conclusion, the paper argues that exploring such assumptions could be a path for developing more relevant HTA approaches for assessing e-health.

B. Health Technology Assessment and RCT

HTA is a research field defined and explained as follows: "the systematic evaluation of the properties and effects of a health technology, addressing the direct and intended effects of this technology, as well as its indirect and unintended consequences, and aimed mainly at informing decision making regarding health technologies. HTA is conducted by interdisciplinary groups that use explicit analytical frameworks drawing on a variety of methods" [8].

The purpose of HTA is to establish a decision basis for procuring the right health technologies [9]. The European network for HTA, EUnetHTA, justifies the research field as follows: "Health care decision making requires the right evidence at the right time. Every day there are new health technologies available that can improve patient outcomes and refine health system efficiency. HTA is a tool to review technologies and provide evidence of the value these technologies can deliver to patients and their families, health system stakeholders, and to society more broadly" [8].

Health technologies comprise "Diagnostic and treatment methods, medical equipment, pharmaceuticals, rehabilitation and prevention methods, but also organisational and support systems used to deliver healthcare" [10]. The inclusion of the EHR as a health technology refers to the description above. The EHR comprises technology but also involves organizational structures, routines, and coordination components. In addition, the EHR includes vulnerable information, ethical considerations, social interactions, relationships, and competencies among users (e.g., doctors, nurses, and patients) which are necessary to deliver health care. In that respect, the EHR can be subsumed under the concept of a health technology, albeit a complex one.

In HTA, different products have been developed to support knowledge-based decisions in health care, such as systematic reviews, meta-analyses, economic modeling, and case or experimental assessments of new medical methods. These tools draw on basic philosophical assumptions and form a coherent approach. The gold standard tool for assessments is RCTs. An RCT is a type of scientific (often medical) experiment, where the people studied are randomly allocated to a treatment or intervention under study. RCTs are often used to test the efficacy or effectiveness of various types of medical interventions, such as drugs, or devices, such as pacemakers. The clinical quality of medications, for example, is also tested via RCTs. The interventions tested for different outcomes are assumed to be clearly defined and demarcated and capable of providing evidence for adverse effects, such as drug reactions.

C. Assumptions and Weaknesses

Accurate assumptions to guide approaches are imperative to produce useful knowledge for different stakeholder groups. The assumption underlying an RCT is that there is a singular reality amenable to objective scientific measurement to provide universal evidence for the outcome of the specified interventions. Human bodies, for instance, are assumed to mainly react similarly to a drug, so that general conclusions can be drawn about its effects. A stable situation is assumed and set up for the experiment, so that causal variables and links can be identified to generalize and repeat the outcome. One challenge for applying RCTs for ICT and e-health programs is that the empirical situation, reality, in general is more messy and in flux [11].

An example is the assessment of electronic health records in the NHS IT program in the United Kingdom. Greenhalgh and colleagues addressed this challenge, and they asserted that e-health initiatives occur in complex and fast-moving socio-political arenas. Evidence is produced by and fed back into a political process of deciding priorities and allocating the resources to pursue them [12]. The authors suggested that interpreting practice in context, therefore, could be an alternative to producing evidence for universal truths in controlled experiments as recommended in RCTs.

A second assumption underlying RCTs is a clear demarcation and definition of the intervention, including a fixed start and endpoint. In ICT programs, this can be difficult to achieve given the fast-paced technological development and the seemingly endless range of possibilities for novel service delivery platforms. It normally takes years to conduct an RCT, and this is described as the most formidable challenge threatening to upset the very promise of potential solutions: The rate of emerging technologies and services far outpaces the field's capacity to demonstrate the conceptual or empirical benefits [13].

A different challenge is the pressure to roll out new ICT services before pilots are fully evaluated. Implementation, thus, is assumed to be a linear operation where readymade technological applications are rolled out to an organization and can be objectively assessed. Human interaction might be considered an obstacle in such processes. The alternative is proposals to address closer person-to-person interaction between users and designers to understand how collegiate and interpersonal elements of care delivery can be better embodied in assessments and therefore brought to consciousness to influence development [14]. In design, the emerging openEHR standard represents a step in this direction as clinical personnel can define for themselves how the content of an EHR should look, that is, the type and degree of various structured elements in order to lay the foundation for interoperability, decision support, and clinical research [6]. OpenEHR is promoted through the international openEHR foundation (a not-for-profit company), and the openEHR standard represents the specification of an EHR system: the management, storage, and retrieval of health data [15].

The three challenges described, and the assumptions behind them of a singular and stable reality, a fixed intervention, and a linear process of implementation, are interconnected. These assumptions and subsequent approaches could fail as guiding principles for addressing all the important aspects that affect knowledge about the value of ICT. Evidence of positive or negative effects based on erroneous assumptions might support overly optimistic and overly pessimistic expectations for future development.

In HTA assessments, different models are, however, defined for the assessment of innovations and form part of the initiatives taken for novel approaches. The Core model distinguishes two models: the diffusion model and the translation model which are relevant to the discussion in the paper. These will be described and discussed in Section III D.

In the remainder of the paper, steps to address these challenges connected to the FIKS program are discussed. The research question is specified and discussed in three parts: How are assumptions about the reality or context, the intervention, the process of implementation, and subsequent approaches and methods of RCT amenable to empirical features of FIKS? Based upon scientific literature, complementary assumptions that can improve HTA for ICT are presented.

II. The Fiks Program, Methods, and Materials

A. FIKS

FIKS is a large-scale program for developing and implementing a new electronic health record system, running from 2012 through 2016. The costs are estimated at EUR 90 million, and the vendor (DIPS) is the largest EHR vendor in Norway [7]. The aim is to introduce a single electronic patient record at the 11 northern Norwegian hospitals, including radiology, lab, pathology, and electronic requisition of laboratory services for general practices in the region [16].

An important goal for the Regional Health Authority, North Norway was to acquire a process- and decision-supportive EHR. Thus, the bid for tender asked for an EHR with high interoperability and configurability that would enable users to tailor the software to their needs. DIPS was commissioned to develop the new EHR infrastructure based on the openEHR architecture [16]. Due to the high configurability associated with an openEHR-based EHR, it was expected that openEHR would have the potential to support collaboration and workflow of flexible patient pathway processes across department and institutional boundaries.

B. Methods and Materials

The paper is based upon mixed data material consisting of documents, web sites, information from advisors, and presentations of the FIKS program to different actors in the hospitals in North Norway over a period of 5 years, from 2011 to February 2016. Data were collected via observations, interviews, participation in meetings and conversations, and access to documents and presentations. The authors collected data, as did PhD students who conducted participatory projects of aspects of the development process. References [6][21][23] and [25] are papers for which PhD students are the first author. The authors and the students are members of the research group of Telemedicine and e-health at UIT - The Arctic University of Norway. Different aspects of the processes have been extensively discussed in the research group, as well as at the Norwegian Centre for e-health Research (NSE). Numerous professionals were involved in these discussions, whom we thank.

In addition, papers and reports from two large-scale evaluation and assessment projects in the UK connected to the NHS ICT program were studied: "The UK Summary Care Record Programme" [12] and "Healthcare Electronic Records in Organisations" [18]. Many scientific papers were recommended in publications from the two programs that focus on assessment traditions. The papers are discussed in the background and discussion sections.

The different data sources are combined through a triangulation process. Triangulation is a social science technique that facilitates validation of data through cross-verification from two or more sources [19]. In particular, triangulation refers to the application and combination of several research methods in the study of the same phenomenon. Such techniques were applied to combine information from multiple sources refined into useable assemblages. These culminated to form recognizable examples for the discussion of assumptions and approaches. The discussion sections also draw on arguments developed with the support of the MethoTelemed team, whose contribution is acknowledged [20].

III. Results and Discussion

A. Assumption One: A Singular Reality Amenable to Scientific Measurement and Control

In this section, the context of FIKS is substantiated, and the program itself can be understood in terms of complexity, multiplicity, and dynamism. Making a clear distinction between the context and the program is not straightforward. Stable variables depicting the reality, or context, are, however, also distinguished and discussed. These are the terms in RCTs for distinguishing external and internal causal variables and links in order to be able to repeat outcomes in controlled ways.

The context of FIKS is many mutually dependent actors, representing numerous interests trying to accomplish a unified vision. This was apparent in the formative stages of the project where several development tracks were established, and more than 150 users were invited to associated regional workshops to identify what needed to be developed. The workshops became an arena for the users from the different hospitals to understand how the practices differed. Accordingly, the users had to negotiate and compromise in order to agree upon standards and trajectories across organizational boundaries. To illustrate, in the first workshops, much time was spent discussing the different needs of small and large hospitals related to the role of coordinator in the operating rooms. In the smaller local hospitals with two rooms, this role had a very different meaning for users than it had for users from a university hospital with 16 operating rooms. This difference had clear implications for designing the new surgical planning module where clinicians from the smaller hospitals generally preferred simplicity in use instead of the more complex functionality that typically was needed at a university hospital.

Another illustration is how differently the vendor's developers and users understood surgery planning. While the vendor planned to develop a functionality in a business-like manner where activities could close at different steps of the surgical planning process to exploit the surgical resources better (personnel and operating rooms), the users argued that planning surgery was a continuous process that did not stop until just a few days before the surgery. Thus, the regional workshops had to take into account different contexts of patient care and treatments in the hospitals.

In this project, the vendor and the hospitals were committed to developing an EHR with highly structured content. Unfortunately, they had different understandings of how this should be achieved. Many users believed that they would get a structured record as part of the delivery of the new EHR. However, DIPS had in accordance with the openEHR architecture made it technically possible for the users to easily do this themselves. Accordingly, in the first pilots, the users missed the structured content. As a result, a national organization had to be established to standardize the EHR.

In addition, the different hospitals where the implementation occurred represent different socio-political and institutional contexts. The context, therefore, is complex, interconnected, and politicized, as health political decisions affect resources necessary to add affordances of, and accommodation of the record. For example, in one aspect it was quite easy to agree to centralize the IT portfolio in a regional version from the previous 11 hospitals. However, an emerging crucial question was who should run and control the daily operation (the regional ICT governance) and who should decide the functionality and use patterns. The university hospital took for granted that it should have this responsibility, but some of the smaller hospitals found this unacceptable. As a solution, in 2014 the University Hospital of North Norway (UNN) suggested a fragmented governance model in which each health trust· meaning each of the eleven hospitals in the region, should be responsible for regulating areas of the ICT portfolio on behalf of the others, meaning that one health trust would govern the EHR: One would regulate the laboratories, another radiology, and so on [21].

In addition, the historical process accounts for the dynamic and interwoven characteristic of the context and intervention. In 2011, contracts for the program had been signed, showing the different industrial actors involved. The web page presented some of the milestones: In 2011, Helse Nord Regional Trust signed a contract with Sectra, Tieto, and DIPS. In 2013, Helse Nord Trust signed a contract with CompuGroup Medical Norge (CGM), and Infodoc. In addition, the pathology systems in two major hospitals merged. In 2015, Helse Nord Trust signed a contract with Hove Medical Systems [7]. The assumptions for the conduct of an RCT, a controlled, measurable, and relatively stable reality, are not reflected in the empirical features of the context/intervention. Instead, the multiple and mutually dependent actors and interests depict a reality under development and flux, depending on negotiations, shifting political conditions, and resources.

B. *Assumption Two: A Clear Demarcation and Definition of the Intervention*

On the FIKS web page and Facebook page, the goal is described as the ambition that the people of the north will have their clinical history assembled in one patient record and that the practice of sending records between hospitals will end. An ambition refers to a work process, not to a defined and fixed intervention as assumed in RCTs. The notion rather refers to assumptions of a creative process as in "Design Thinking". One of the first features of the intervention was described in 2013: "Moving the databases of the hospitals in health region to one central common database, is an important condition for the implementation of common patient administration and treatment systems and one common electronic record for the individual hospitals in North Norway" [7].

New components were added to the service. Events were planned as an ongoing deployment process, and the program was described in terms of technical and operational events: the connection of the Narvik medical center to the Health Nords regional solution for electronic requisitioning of laboratory services (11/25/2015), the connection of the Leirfjord medical office to the Health Nord regional solution for electronic requisitioning of laboratory services (11/25/2015), the connection of the Nordreisa medical office to the Health Nords regional solution for electronic requisitioning of laboratory services (11/24/2015), the connection of Træna medical office to the Health Nords regional solution for electronic requisitioning of laboratory services 11/24/2015), and the connection of the Skjervøy medical office to the Health Nords regional solution for electronic requisitioning of laboratory services [22].

By distinguishing events this way, the foundation is laid for RCTs of each part of the process, but the resources needed for this endeavor would be vast. There are also connections between the parts, and therefore, it is difficult to single out clearly demarcated interventions. For example, the different hospitals had older versions of DIPS (DIPS classic) in play, as well as existing systems for other areas, such as the laboratory and radiology. The implementation of the new portfolio cannot be achieved without taking into account foreseen and unforeseen constraints (and opportunities) in the existing systems.

Based on this, the senior management found it was enormously risky to replace the existing portfolio in one stroke and decided to pursue a stepwise strategy in which different modules of the existing portfolio were replaced at a time [23]. This made it essential that the new system had to be compatible (data integration) with the old system since they were supposed to work together over several years. Since this also was a development project, it was easy for the users to give feedback to the vendor on what worked and what did not and how newer versions of the software should be tailored to the users' need. In this way, the implementation was far from clear-cut but had to be adapted based on existing practice, functionality in existing technology, and users' feedback.

Moreover, to make the software flexible to allow future configurations, the EHR was developed in accordance with the openEHR approach, which encourages users to make changes to the software themselves. In this way, the EHR will never be a finished product but will continue to grow and transform after it has been put into daily operation.

Another process that affected the implementation of the new EHR was another large-scale project that started in 2014. The goal of the project is to implement an electronic medication management (EMMS) system in the health region [24]. The EMMS and the new EHR are supposed to be closely integrated, meaning that many adaptions have to be made in both systems on what to integrate, when to integrate, and how to integrate them. Since the EHR and the EMMS are crucial systems for clinicians, and have much overlapping functionality, it is not obvious in both projects how to establish the integration.

Currently, there is an ongoing discussion about which one of the two systems that should archive the master data (true version) of overlapping and integrated data. One example is information that a patient is allergic to certain medications or food, which is crucial information for the EMMS and the EHR.

When it comes to the affordances of FIKS, the web page states that the next-generation patient record is under development and is being tested in the region. Some milestones on the path to one common medical record were listed in 2016: There is one common medical record at the hospitals in Hammerfest and Kirkenes, the UNN employs regional radiology solutions, including a common radiology archive, and the hospitals in Helgeland, Mo I Rana, Mosjøen, and Sandnessjøen employ one common medical record (DIPS) [7].

This information tells us that the contexts and the intervention consist of multiple, developing, and mutually dependent components and processes.

C. *Assumption Three: Implementation as a Linear and Controlled Operation*

The notion "roll out new ICT services before pilots are fully evaluated" involves the assumption that implementation is a linear, top-down, and controlled process of a ready-made intervention. In addition, it involves an assumption that implementation can be distinguished from context and socio-political or human processes. The description of the implementation of FIKS, however, clearly points to an ongoing and changing process where different components should be aligned. This is how it is expressed on Facebook: The FIKS program in Helse Nord consists of six projects intended to develop and implement joint electronic record systems at the hospitals in Northern Norway: one joint electronic record (DIPS), new features of the electronic record (DIPS Arena), a laboratory information system, radiology systems (Sectra, RIS, and PACS), joint pathology system in Tromsø and Bodø, and electronic requisition of laboratory services [22].

In 2011, DIPS decided to use the openEHR framework to develop its next-generation EHR for the hospital market [5]. This involves negotiations on development directions. The role of interaction between different participants in the process is a collegial and interpersonal process, enacted as different meetings for dialogue and negotiations: 11.26.2015, Workshop (EHR Development), Theater nurse meeting, Planning and booking DIPS Arena; 11.26.2015, Workshop (EHR Development), Theater nurse meeting, meeting with clinicians; 11/26/2015, Workshop (EHR Development), Decisions in psychiatry, new module in DIPS Arena; and 11/12/2015: Operation Planning (EHR Development) and meeting with clinicians at University Hospital of North Norway [7]. This process adds to the previously documented process of negotiating contracts with producers and vendors.

FIKS also designates and educates super-users and states that employees' competencies are crucial for the development of good record systems: "Close to 190 super users at Nordlandssykehuset are ready to be educated on use and routines of the new electronic record and become leading DIPS experts" [7].

The description of the implementation shows a multitude of inter-related operational, interactional, and relational processes. Thus far, the roll-out process has been far from linear, pre-defined, and controlled. It has, instead, been characterized by continuous interaction, discussions, and tensions between different users and between users and developers. In turn, this has transformed how the vendor collects requirements for functionality, how new functionality should be tested, and how and whom to include in the various steps of the design and implementation process.

For example, at the beginning of the project, the vendor invited users to define their requirements through user stories that were small descriptions (three to four lines) of work situations. The developers then used these stories as a basis when developing the new functionality. However, this appeared to be very problematic because due to the heterogeneous user group, it was difficult for the vendor to find coherence in the many (and diverging) user stories [25]. As an alternative, the vendor had to change the method for communicating with users.

The users now had to define steps of, for instance, surgery planning in a several-page document, and the process revealed that the steps were performed by different professionals: sub-specialized physicians, nurses, and secretaries. This diversity implied that the developers had to broaden their perspective to look beyond the physician's role. Even so, it became increasingly clear that the developers needed more contextual data, and therefore, they decided to spend more time in the hospital departments in order to identify what was needed of the new EHR in specific work situations [25].

The first pilot of a smaller segment of the new EHR conducted in a smaller hospital in southern Norway also revealed that there was a clear gap between the functionality offered in the system and what was needed in clinical practice. This gap underscored that users had to be even more involved in experimentation and testing of new functionality before it was put into action.

A crucial project management issue was that after four years of development, the UNN stated that it had spent too much user resources on the project and therefore wanted more in return from the vendor. It was difficult to come to an agreement on this, and a hospital in western Norway replaced the university hospital as the key collaborating partner in the development of the new EHR.

A linear, pre-defined, and controlled roll-out process is not present, as assumed in order for an ICT to be performed and produce generalizable knowledge about the effects of an implemented intervention.

D. Summary and Discussion

Among the challenges in applying a HTA framework for the study of effects of an electronic patient record is that HTA tools form a coherent approach and draw on common basic assumptions. As shown in the accounts of FIKS, the assumptions differ from the empirical features. The basic assumptions of a stable reality where generalizable effects of a pre-defined intervention can be transferred to other settings via an operational and linear implementation process, fail to address the empirical features described. This issue has been discussed in different HTA bodies, related to innovation research. As mentioned in the introduction, two different models or approaches are described in the core model. The first of these is the linear diffusion model, which perceives new technology as an external stable entity that is brought to a (health care) system and induces change [26].

A competing paradigm, the translation model is also embedded in the core model. Within this model, technology is perceived to undergo changes as it interacts with the environment in which it is used. Hence, the final impact will not depend on the original technology only [26]. In the case of FIKS, both the technology, the health-care setting or environment, and the implementation process seem to be in a state of mutual translation. The findings, therefore, encourage openness to even more complexity than the two different models denote, in assumptions governing assessments. The empirical features of the applications and services connected to the record, are highly diverse and constantly in flux within the shifting social and organizational contexts.

Challenges were connected to the discrepancies between the RCT assumptions and the features of the context or reality within which the electronic record is embedded, the intervention itself, and the implementation process. In the next section, approaches that build on other assumptions are briefly addressed.

IV. Complementary Assumptions: The Constructivist Umbrella

Challenges concerning the validity of evidence in the face of the involvement of different stakeholders have been articulated within the HTA tradition, which is looking to overcome such challenges. One HTA tool is consensus conferences with different stakeholders [26]. Such conferences have been investigated, and the following assertion strengthens the argument of a shifting social reality and the need to consider social relations as drivers for intended and unintended outcome:

"Consensus development programs are not immune to the economic, political, and social forces that often serve as barriers or threats to evidence-based processes. Organizations that sponsor consensus development conferences may do so because they have certain expectations for the findings of these processes, and may find themselves at odds with evidence-based findings. Other stakeholders, including from industry, biomedical research institutions, health professions, patient groups, and politicians seeking to align themselves with certain groups, may seek to pressure consensus development panelists or even denounce a panel's findings in order to render desired results, the evidence notwithstanding" [27]. This is a long quotation, which we have chosen to cite because it illustrates the fact that HTA institutions are highly up to date on the complexity of evidence production. The importance of addressing social and political interests and processes are clearly recommended in order to understand the way evidence can be produced and affect the results of ICT use in health services.

In contrast to assumptions in RCTs about a stable and objective reality, a fixed intervention, and a linear and controlled implementation process, constructivist traditions assume that flux occurs. This comprise that reality is under development, the interventions are subject to change, and implementation is partly unpredictable and depends on, for instance, resource allocation. Implementation is considered an ongoing process where certain types of support or lack of support strongly influence the outcome. Therefore, it is not considered possible to generalize evidence-based outcomes to repeat good results in new or future settings. Within such traditions, a formative idea exists. It implies to feed assessment results, which are produced along with the development processes, back into a pragmatic and political process of deciding priorities and allocating resources to pursue them.

In this perspective, validation is obtained through negotiations between the context, the researchers, the intervention, and other stakeholders. Context is considered by involving different stakeholders' interests, and validity is addressed by asking what the study is valid for [28].

Such assumptions and resulting approaches may have particular strengths when the goal is to develop good e-health services, to the confidence of users, professionals, policy makers, and payers, and as a leading market in Europe. Thus, obtaining a balance between different validity claims is a huge challenge.

In a paper building on discussions in the annual meeting in 2016 within HTAi, the topic of changing HTA paradigms was addressed [29]. The new HTA paradigm is characterized by a more agile and adaptive HTA process across the life cycle of technologies, reacting quickly to new and real life data when they become available or when changes in the technology life cycle emerge; assessment methods and language that go beyond incremental cost-effectiveness ratios, incorporating meaningful results for clinicians and patients; where information on what the health system and patients need from innovation, and what the health system may need to do to get value from it, is discussed through the lens of health services delivery and product lifecycle; through multi-lateral stakeholder dialog and collaboration that addresses health needs and product conceptualization; through development, evaluation, introduction, and appropriate use in a changing landscape as other developments come on stream.

As the last point in this discussion, we focus on research networks and institutions where assessments and evaluations have been carried out, explored, and substantiated over more than three decades. We briefly introduce several papers and highlight basic assumptions. The point is to highlight shared basic assumptions to comment on a discussion of the need to unite networks to develop more operational approaches to assessments of e-health, as well as to unite forces.

What seems to characterize some of the research networks, is that papers have been produced periodically. Constructive technology assessment (CTA) was introduced in the 1990s [30]. It shifts the focus away from assessing impacts of new technologies to broadening design. This approach in general, assumes that design and assessments are co-producers of innovations, a dynamic and processual view of the context, intervention, and implementation processes [31]. Numerous papers were published around the start of the 1990s. Around 2005, a new wave of papers were published on the subject; see, for instance, Genus, who discusses assumptions of stakeholders and democratic processes as foci important for assessing added value [32]. Since 2013, papers and books have addressed pragmatic evaluations, and different institutions seem to share basic assumptions of dynamics and the use of real life data. Monitoring data is, for instance, a fast-growing option for knowledge production that invites collaboration between different assessment networks and units [33]. This point has been promoted by the Organization for Economic Cooperation and Development (OECD). Scholars have also recently argued for the need to unify the efforts of different evaluation and assessment networks; see, for instance, [34].

What is suggested for a new paradigm in HTA seems to reflect a combination of two broad traditions, the positivist as in RCTs and the constructivist as briefly outlined. To bridge the gaps between the assumptions of the two traditions, the positivist and the constructivist, should be important to produce valuable e-health assessments. This point was also noted by Ammenwerth et al. [35]. One goal should be to open the borders between traditions and identify how evaluators may draw on the benefits the different ones have to offer.

The different networks and institutions briefly introduced in this section are the Science and Technology Studies network (STS) and the Medical Informatics network. Another research environment to consider is various social science networks, which have a long tradition for theoretically advanced assessment approaches. In this short introduction, we mention the book by Guba and Lincoln called "Fourth Generation Evaluation" [36]. The book was followed by a number of discussions on "naturalistic inquiry". Fourth Generation Evaluation moves beyond science to include the myriad human, political, social, cultural, and contextual elements that are involved in processes of change. Based upon relativism, a unity between knower and known, and a subjective epistemology, the authors show how the concept fourth generation evaluation unites the evaluator and the stakeholders in interaction to co-create the product of the evaluation.

All these networks are highly present communities in assessments of e-health and telemedicine. Their work should be considered, as well as initiatives within the OECD domain. We also introduced the MethoTelemed Team [20]. As a follow-up of the work in [20], this team is currently embarking on a review to assess telemedicine and e-health assessments, the methodologies used, as well as the review author's conclusions. This work should provide additional clues about how different traditions have developed concerning assumptions and approaches. This work could strengthen the knowledge base for discussing joint forces.

Answers to the question "Does it work?" to produce evidence for universal truths need to be supplemented by a whole range of answers to questions that reflect the complexity of most e-health interventions: How does it work? Who does it work for? What components are vital to success, and which are redundant? Why does it work in this context (and equally important not work)? Is this an appropriate and acceptable way of tackling the problem? How is quality produced and defined within certain innovation processes? Who owns the definition of success?

This paper focused on the fluctuating character of reality, interventions and implementation processes alon with the development of FIKS. Nevertheless, the three elements have some stable features. In process approaches, investigations are directed toward the conditions included in the development processes, to feed the results back into the process for dialogue and improvements. The intervention is shaped and adjusted in and through practices of professional–social interaction between participants (doctors, nurses, and patients) and the organizational, economic, political, and ideological settings in which these practices are embedded. This approach is formative. The intervention also contributes to shaping these settings as the approach pre-supposes that all entities are in a process of mutual shaping. Influence and success are empirical questions, in turn, potentially enacting different answers in each situation assessed. Obtaining the power to control conditions will be the crucial task for the future results of innovations. Power may be based on participatory or top-down models. Process investigations may produce knowledge to understand how successes are defined, produced and maintained.

V. Conclusion and Further Work

The paper has substantiated empirical features of a reality in flux, an intervention under development, and implementation processes as ongoing negotiations for the FIKS program. HTA assumptions of a stable reality, a fixed intervention and a controlled implementation process were not present. Steps to strengthen HTA use for ICT are timely even if the HTA communities have not been extensively attentive to e-health. E-health communities have also not been attentive or interested in HTA. Steps to unite different research networks and institutions that make profound efforts to address the vast task of assessing ICTs in health care are also timely. Approaches that take into consideration dynamics and complexities in contexts, interventions, and processes of implementation are discussed in all communities. Knowledge about the conditions for large processes with escalating costs is important, as conditions built into the programs vastly influence the effects that emerge and manifest.

Embedding assumptions of a world in flux where social, technical, and clinical entities influence each other in dynamic processes should increase the relevance of HTA of ICT and affect real-time developments. We suggest further exploration of assumptions that encourage participatory and process assessment approaches. Such assumptions will strengthen the knowledge base for future procurements, as human interactions play important roles in the development. We also recommend controlled studies of stable components of e-health.

ACKNOWLEDGMENT

We are grateful to the North Norwegian Health Authorities that funded the project and to communication advisor Elin Vinje Jenssen, who rapidly translated the FIKS fact sheet into English.

REFERENCES

[1] A. G. Ekeland, "Assessing electronic health records: Are basic assumptions in 'Health Technology Assessment' useful?" eTELEMED 2016, IARIA, pp. 36–41, April 2016.

[2] The Parliament, U.K., The National Programme for IT in the *NHS: An Update on the Delivery of Detailed Care Records Systems, HC (2010-12)*. [Online]. Available from: http://www.publications.parliament.uk/pa/cm201012/cmselect/cmpubacc/1070/107003.htm 2016.03.14

[3] International Network of Agencies for Health Technology Assessment (INAHTA), Health Technology Assessment international (HTAi) and Other Partner Organizations, *HTA Glossary*. [Online]. Available from: http://htaglossary.net/HomePage 2016.03.14

[4] A. G. Ekeland, *Health Technology Assessment (HTA) for ICT: Development and Adaptation*. [Online]. Available from https://ehealthresearch.no/en/projects/health-technology-assessment 2017.04.12

[5] The Norwegian Government, *One Inhabitant - One Record, Meld. St. 9 (2012–2013)*. [Online]. Available from: https://www.regjeringen.no/no/dokumenter/meld-st-9-20122013/id708609/?q=Meld 2016.03.14

[6] B. Christensen and G. Ellingsen, *"Evaluating model-driven development for large-scale EHRs through the openEHR approach,"* Inter. J. Med. Inform., vol. 89, pp. 43–54, May 2016.

[7] Helse Nord, *Implementation of common clinical ICT systems (FIKS)*. [Online]. Available from: http://www.helse-nord.no/about-fiks-in-english/implementation-of-common-clinical-ict-systems-fiks-article135571-42080.html 2016.03.14

[8] Eunethta, *Mission, Vision & Values*. [Online]. Available from: http://www.eunethta.eu/about-us/mission-vision-values

[9] B. Jennett, "Health technology assessment," BMJ: Brit. Med. J., vol. 305, pp. 67–68, 1992.07.11.

[10] Eunethta, *HTA Common Questions*. [Online]. Available from: http://www.eunethta.eu/about-us/faq#t287n73 2016.03.14

[11] T. Scott, T. G. Rundall, T. M. Vogt, and J. Hsu, "Implementing an Electronic Medical Record System: Successes, Failures, Lessons," Oxford: Radcliffe, 2007.

[12] T. Greenhalgh and J. Russell, "Why do evaluations of eHealth programs fail? An alternative set of guiding principles," PLoS Med, vol. 7, pp. e1000360, 2010, doi:10.1371/journal.pmed.1000360

[13] D. J. Jones, "Future directions in the design, development, and investigation of technology as a service delivery vehicle," J. Clin. Child Adolesc. Psychol., vol. 43, pp. 128–142, January 2014.

[14] D. Avison and T. Young, "Time to rethink health care and ICT?" ACM, vol. 50, pp. 69–74, June 2007, doi:10.1145/1247001.1247008

[15] OpenEHR, *What is openEHR?* [Online]. Available from: http://www.openehr.org/what_is_openehr 2016.03.14

[16] S. Kallbekken, *The first step on the road to one patient record in Helse Nord (Første steg på veien mot en pasientjournal i Helse Nord.)* [Online]. Available from: http://www.unn.no/aktuelt/forste-steg-pa-veien-mot-en-pasientjournal-i-helse-nord-article108606-10106.html 2016.03.15

[17] T. Beale and S. Heard, "An ontology-based model of clinical information," in Medinfo 2007: Proceedings of the 12th World Congress on Health (Medical) Informatics; Building Sustainable Health Systems. Amsterdam: IOS Press, pp.760-765, 2007.

[18] Healthcare Electronic Records in Organisations, *Research Perspectives Facts and Figures*. [Online]. Available from: http://www.researchperspectives.org/ 2016.03.15

[19] A. K. Bekhet and J. A. Zauszniewski, "Methodological triangulation: an approach to understanding data," Nurse Res., vol. 20, pp. 40–43, 2013.01.16 doi 10.7748/nr2012.11.20.2.40.c9442.

[20] A. G. Ekeland, A. Bowes, and S. Flottorp, "Methodologies for assessing telemedicine: a systematic review of reviews," Int. J. Med. Inform., vol. 81, pp. 1–11, November 2011.

[21] G.-H. Ulriksen, R. Pedersen, and G. Ellingsen, "Establishing ICT governance for regional information infrastructures in healthcare," 49th Hawaii International Conference on System Sciences (HICSS), pp. 5137–5146, 2016.

[22] FIKS. Helse Nord: *FIKS*. [Online]. Available from: https://www.facebook.com/HelseNordFIKS/info/?tab=page_info 2017.05.24

[23] L. Silsand and G. Ellingsen, "Generification by translation: Designing generic systems in context of the local," J. Assoc. Inf. Syst., vol. 15, pp. 177–196, April 2014.

[24] J. I. Westbrook, L. Li, A. Georgiou, R. Paoloni, and J. Cullen, "Impact of an electronic medication management system on hospital doctors' and nurses' work: a controlled pre-post, time and motion study," JAMIA, vol. 20, pp. 1150–1158, November 2013.

[25] B. Christensen, L. Silsand, R. Wynn, and G. Ellingsen, "The biography of participation," in Participatory Design Conference (PDC) '14. Windhoek, Namibia: Proceedings of the 13th Participatory Design Conference: Short Papers, Industry Cases, Workshop Descriptions, Doctoral Consortium papers, and Keynote abstracts - Volume 2, pp. 1–4, 2014.

[26] Eunethta, *HTA Core Model for Medical and Surgical Interventions.* [Online]. Available from: http://meka.thl.fi/htacore/model/HTA%20Core%20Model%20for%20Medical%20and%20Surgical%20Interventions%201.0r.pdf 2016.03.15

[27] C. Goodman, *Introduction to Health Technology Assessment.* [Online]. Available from: http://aaz.hr/resources/pages/57/2.%20HTA101-%20Introduction%20to%20health%20technology%20assessment.pdf 2016.03.15

[28] J. Aguinaldo, "Rethinking validity in qualitative research from a social constructionist perspective: From 'Is this valid research?' to 'What is this research valid for?'" Qual. Rep., vol. 9, pp. 127–136, March 2004.

[29] D. Husereau, C. Henshall, L. Sampietro-Colom, and S. Thomas, "Changing health technology assessment paradigms?" Int. J. Technol. Assess. Health Care, pp. 1–9, January 2016.

[30] J. Schot, "Constructive technology assessment and technology dynamics: The case of clean technologies," Sci. Technol. Hum. Values, vol. 17, p. 36, January 1992.

[31] A. Rip and H. Van den Belt, Constructive Technology Assessment: Toward a Theory. Enschede: University of Twente, 1988.

[32] A. Genus, "Rethinking constructive technology assessment as democratic, reflective, discourse," Technol. Forecast. Soc. Change, vol. 73, pp. 13–26, 2006

[33] H. Hyppönen, A. Faxvaag, H. Gilstad, G. A. Hardardottir, L. Jerlvall, M. Kangas, S. Koch, C. Nøhr, T. Pehrsson, J. Reponen, Å . Walldius, and V. Vimarlund, "Nordic eHealth Indicators: Organisation of research, first results and plan for the future," Stud. Health Technol. Inform., vol. 192, pp. 273–277, 2013.

[34]P. Doupi, "Evolving health IT systems evaluation: the convergence of health informatics and HTA," in E. Ammenwerth and M. Rigby, Eds. Evidence Based Health Informatics. Amsterdam: IOS Press, pp. 220 - 236, 2016.

[35] E. Ammenwerth, J. Brender, P. Nykänen, H. U. Prokosch, M. Rigby, J. Talmon, and HIS-EVAL Workshop Participants, "Visions and strategies to improve evaluation of health information systems. Reflections and lessons based on the HIS-EVAL workshop in Innsbruck," Int. J. Med. Inform., vol. 73, pp. 479–491, June 2004.

[36] E. G. Guba and Y. Lincoln, "Fourth Generation Evaluation," SAGE 1989

Becoming a Medical Device Software Supplier and Complying with Data Security Regulations

Fergal McCaffery, Ronald Jabangwe, Kitija Trektere,
Regulated Software Research Centre & Lero,
Dundalk Institute of Technology
Dundalk, Co.Louth, Ireland
e-mail: {fergal.mccaffery, ronald.jabangwe, kitija.trektere }@dkit.ie

Garret Coady
BlueBridge Technologies
3015 Lake Drive, Citywest, Dublin 24, Ireland
e-mail: {garretcoady}@bluebridgetech.com

***Abstract*—Today many software development companies are restructuring their business model to enter the medical device domain. The reason for this change is that significant opportunities exist within the healthcare industry and particularly in relation to the usage of software within this domain. However, in order to become either a medical device software supplier or manufacturer there are challenges to overcome, and data protection regulations to abide by. This paper describes a case study of an Irish software development company that in 2014 decided to change their business model to enable them to become a medical device software supplier, and engaging with clients in the United States of America. The paper provides an account of their journey from being an automotive software supplier to securing software development contracts from leading medical device manufacturers. This involved them having to re-design and re-structure their software development approach to meet both the demands of medical device standards, data security regulations and medical device multinational third party software selection criteria.**

Keywords-MDevSPICE® Framework; Software Development Process; Medical Device Software; Software Security; HIPAA; Agile Software Development.

I. INTRODUCTION

The enormous and seemingly ever-growing medical device market value motivated the case company presented in this paper to shift to a medical device software supplier [1]. In 2015, the medical device (MD) global market was "valued at $228 billion, up from $164 in 2010 and projected to reach $440 billion by 2018" "at approximately 4.4% compound annual growth rate per year" [2]. The leaders in the MD market are the United States of America (USA) having 38% of the global value of this market followed by China with a market valued at $48 billion with western Europe having almost 25% of the global market [2]. However, to become a MD supplier for the industry takes significant time and resources as there are many obstacles that need to be overcome.

This paper extends the paper presented in [1], which is based on a case study of an Irish software development company *BlueBridge Technologies* (BBT). Their journey started in 2014 when BBT decided to embark upon becoming a MD software supplier and at that moment they had no regulatory requirements in place, in fact a key question they asked at that stage was *"what are the standards we need to implement and in what order?"*. This paper presents how with the help of academic MD researchers' regulations were put in place through undergoing an MDevSPICE® assessment and outlining the challenges that might arise in the near future.

The rest of this paper is organized as follows. Section II describes the background of BBT and the current situation in the MD industry. Section III outlines the challenges BBT faced in order to become a MD software supplier. Section IV describes the approaches followed to become a MD software supplier. Section V outlines given that BBT have satisfied the regulations they wish to further refine and improve their software development processes to make them more efficient. Section VI describes first steps taken in order to improve their current lifecycle process and approach to ensuring data security. Recommendations to the case company are provided in Section VII. The final section of the paper provides a conclusion and future work in Section VIII.

II. BACKGROUND OF THE COMPANY

BBT was founded in 2006 – initially formed upon the closure of the Irish based development operations of Magna Automotive, and today employs 19 people with 8 of them working as software developers. BBT are currently working on 7 different projects with 5 of them involving developing the software component for another organization's product. Their current customers include pharmaceutical and multinational MD companies.

The main reason why software development companies wish to enter the MD domain is because of the expansion of the MD industry in the past few years therefore providing many opportunities for others to enter into this industry. The MD industry is largely research and development driven.

Software increasingly performs an essential role in the provision of healthcare services [3]. This is particularly reflected in the importance that software now plays in medical diagnoses and treatment [4]. The level of software functionality in MDs and the complexity of that software has substantially increased [5]. The MD regulatory environment has been extended to include more focus on software. For example, the latest amendment to the Medical

Device Directive [6] recognizes that standalone software can be classified as a MD in its own right. Consequently, a significantly increased proportion of software applications will now be classified as MDs and must be developed in a regulatory compliant manner [7].

Medical records are increasingly being stored in electronic form. The use of Electronic Medical Record (EMR) systems in the USA by physicians increased from 18.2% in 2001 to 48.3% in 2010 [8]. The adoption of EMR systems could produce efficiency and safety savings of $81 billion annually and improve prevention of medical diseases [9]. Use of Mobile devices in health care is increasing. "By 2017, mobile technology will be a key enabler of healthcare delivery reaching every corner of the globe" [10].

III. Challenges BBT Needed to Overcome to Become a MD Software Supplier

To become a MD software supplier there were regulations and standards that needed to be adhered to. This required processes to be defined in accordance with these standards and regulations and then for objective evidence to be obtained demonstrating the implementation of the defined processes. For BBT, the starting point was to gain an understanding of three main standards and data protection regulations in the US. The paragraph below briefly outlines the standards that BBT familiarized themselves with before starting to define their MD software development processes.

A. ISO 13485:2006

"This International Standard specifies requirements for a quality management system that can be used by an organization for the design and development, production, installation and servicing of medical devices, and the design, development, and provision of related services" [11].

ISO 13485 is in practice required by any MD company. It details the requirements for the Quality Management System (QMS) for MDs. The standard is broadly based on ISO 9001, although the 2015 revision of the latter departs significantly from the previous approach. ISO 13485 was recently revised in 2016, resulting in a better alignment with the Food and Drug Administration (FDA) regulations; changes include explicit requirements for validation of software infrastructures used by the company.

ISO 13485 is a "Harmonized Standard" for the EU and a "General Consensus Standard" for the FDA.

Certification to ISO 13485 is achieved by independent audit by a Notified Body of the Quality System of the company. It involves yearly surveillance audits and re-certification every 3 years.

B. IEC 62304

"This standard defines the life cycle requirements for MEDICAL DEVICE SOFTWARE. The set of PROCESSES, ACTIVITIES, and TASKS described in this standard establishes a common framework for MEDICAL DEVICE SOFTWARE life cycle PROCESSES" [11].

IEC 62304 covers the development process for medical device software. This standard is harmonised with the requirements of ISO 13485 and therefore complements it by adding the specifics required for MD software.

Similarly to ISO 13485, IEC 62304 is a "Harmonized Standard" for the EU and a "General Consensus Standard" for the FDA.

However, IEC 62304 interfaces with ISO 13485 in two areas: software inputs and system integration. The software inputs are generated from the system (or subsystem) level requirements, while IEC 62304 explicitly does not cover system level activities, in particular design validation.

Although this is the gold standard for the development of MD software, there is no such thing as accreditation or certification to IEC 62304. Anyway company can request an "independent certification" by a Notified Body and this is particularly attractive to MD software suppliers.

C. ISO 14971:2009

"This International Standard was developed specifically for medical device/system manufacturers using established principles of risk management. For other manufacturers, e.g., in other healthcare industries, this International Standard could be used as informative guidance in developing and maintaining a risk management system and process"[13]. "This International Standard deals with processes for managing risks, primarily to the patient, but also to the operator, other persons, other equipment and the environment" [13].

ISO 14971 is particularly important to any MD manufacturer and supplier. Most decisions made during the whole lifecycle of a device must be risk-based.

The area of regulatory standards and the recording of documentation associated with their implementation was new to BBT. Therefore, BBT engaged with both standards consultants and an academic research group (the RSRC, our research centre) specializing in MD software development research. This assisted BBT to fast-track the initial steps to becoming a MD software supplier.

D. Data Security Regulations: Protecting Health Information

In the US, the law that outlines and standardizes the protection of health information is HIPAA (Health Insurance Portability and Accountability Act) [29]. HIPAA refers to health information as protected health information (PHI) or electronic protected health information (EPHI).

PHI or EPHI includes health information and any accompanying information that can be used to identify an individual, such as, demographical information, that is created, stored, transmitted or maintained when providing health-related services [30].

HIPAA mainly consists of four Rules that state what and how data should be protected, which are [29]: the Privacy Rule, Enforcement Rule and the Breach Notification Rule, Security Rule. The Privacy Rule outlines the standard expected for the protection of personal health or medical information. In the event of violations, for example, inappropriately protecting PHI, the provisions on penalties and related procedures are outlined in the Enforcement Rule. Provisions for the required course of action in case of any breach on protected health information is outlined in the Breach Notification Rule. The Security Rule outlines safeguards that are needed to secure EPHI/PHI that is either created, stored or transmitted.

IV. Approach to Become a MD Software Company

When BBT reached out to the RSRC, we knew that this was an ideal company to become involved with in regards to performing research into how software companies could make the transition to becoming MD software suppliers.

A. Embark on MDevSpice® assessment

First of all, it was essential to understand BBT's current position in regards to their software development processes. We decided to perform an MDevSPICE® [14] assessment. MDevSPICE® is a framework assessment model where all MD software standards and processes are brought together into one place with software engineering best practices. MDevSPICE® was developed in the RSRC. Then, this framework assessment model was utilized in BBT to assess the current situation.

Below we describe what happened next in regards to both the assessment and BBT's subsequent journey to becoming a MD software supplier.

A) Assessment conducted: Given that MDevSPICE® consists of 23 processes we selected the most appropriate 10 processes from the MDevSPICE® model to assess BBT against (see Table I).

It was agreed upon discussion with BBT that only the most foundational processes would be assessed. Therefore, the following 10 out of the 23 MDevSPICE® processes were chosen to be assessed over 2 onsite days in BBT.

The order of the processes assessed was important as it is important to follow the medical device software development lifecycle. Therefore, systems requirements were a very natural place to start. Below is outlined the process assessment schedule: we assessed 5 processes on each day (see Table II).

Each process was assessed by 2 MDevSPICE® assessors in an interview with at least 2 members of BBT being present in each interview. Prior to the interviews both the schedule and the names of the BBT staff members that would be involved in each process interview was agreed. It was very important to ensure that access was provided to the most relevant staff for each interview session as otherwise the assessment would not have been as accurate as possible.

TABLE I. PROCESSES OF MDEVSPICE®

MD System Lifecycle Processes	*MD Software Lifecycle Processes*	*MD Support Processes*
Project Planning	Software Dev. Planning	Configuration Management
Project Assessment and Control	Software Req. Analysis	Software Release
Risk Mgmt. Stakeholder	Software Architectural Design	Software Problem Resolution
Req. Definition	Software Detailed Design	Software Change Request Management
System Req. Analysis	Software Unit Implementation. and Verification	Software Maintenance
System Architectural Design	Software Integration and Integration Testing	
System Integration	Software System Testing	
System Qualification Testing	Software Risk Mgmt.	
Software Installation		
Software Acceptance Support		

TABLE II. DAY 1 AND DAY 2 OF ASSESSMENT PROCESS

Onsite Assessment Day 1
System Requirements Analysis
Software Development Planning
Software Requirements Analysis
Software Architectural Design
Software Detailed Design
Onsite Assessment Day 2
Software Unit Implementation & Verification
Software Integration & Integration Testing
Software System Testing
Software Risk Management
Software Configuration Management

Each of the 10 interviews lasted approximately 1 hour and involved one assessor asking BBT staff a set of scripted questions related to that process area. The second assessor used a tool to record detailed responses from the interviewees with both assessors using the tool to enable each question to be scored as "Fully Achieved", "Partially Achieved" or "Not Achieved". In addition to the usage of predefined scripted questions additional questions were also asked that were specific to BBT.

B) Findings produced: The MDevSPICE® assessors at the end of Day 2 returned back to the RSRC and went through each process together, discusssing the observations and notes from the assessment. As a result of performing the assessment we provided BBT with a set of strengths, issues and recommendations to address those issues across each of the assessed processes. The MDevSPICE® assessment provided coverage over a number of different MD software related standards. Figure 1 shows a breakdown of the coverage provided for each of the different standards from assessing 10 of the 23 MDevSPICE® processes. As one of the goals of BBT Management was to gain an understanding in relation to the state of their current development processes against IEC 62304, as this is the main MD software process standard, processes were selected from MDevSPICE® that featured heavily in IEC 62304. The exception to this was System Requirements Analysis but this was deemed to be a critical process to examine as BBT would be performing software development for an overall MD system. Therefore, it is essential that they have an efficient process in place for System Requirements Analysis as otherwise everything that occurs afterwards within the development lifecycle will be impacted.

From looking at Figure 1 it can be seen that the 10 processes assessed provided: 59% coverage of IEC 62304; 2% of ISO 80002-1 [15] (this technical report relates to how ISO 14971 may be applied within software); 16% of the FDA's Guidance for off the shelf software [16]; 1% of the FDA's Guidance for premarket submissions [17]; 20% of the FDA's Guidance for validation of software [18]; 1% of ISO 13485 and 1% of software engineering best practice standards.

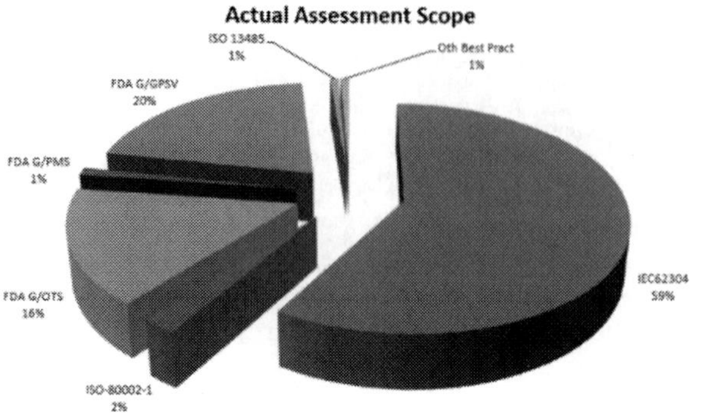

Figure 1. Scope of the BBT Assessment

C) Implementing the recommendations: In order to assist BBT to implement the recommendations in a timely manner BBT took the following steps:

a) Brought in consultants to assist with the implementation of QMS 13485.

b) Recruited an engineer from a leading MD manufacturer who possessed considerable experience in developing MD software development and in particular MD risk management expertise to put in place a risk management strategy in line with ISO 14971.

c) Engaged with a notified body organisation to prepare them for an official audit in IEC 62304 and subsequently perform the audit. This enabled an successful IEC 62304 audit to be achieved in a timely manner.

D) Actions taken by BBT:

a) Gained Certification to ISO 13485

b) Gained Independent Certification in IEC 62304.

c) The IEC 62304 audit was performed against one project using a plan driven approach.

d) BBT have the MD standards that MDs maufacturers request software suppliers to have in place.

The main criteria MD manufacturers use for selecting a MD software supplier is that organizations should have IEC 62304 in place. At this stage BBT now have not only satisfied this criteria but surpassed it in that they not only adopted IEC 62304 but were certified against it and also have adopted IEC 13485, ISO 14971, 21 CFR 820 and the FDA Guidance documentation for: Off the shelf software, Premarket Submissions and Validation of MD software. Therefore, at this stage BBT were ready to obtain contracts as a MD software supplier company.

E) Addressing HIPAA Safeguards

We are currently working with BBT to ensure that their software meets HIPAA requirements. In order to so, we are using the international medical device security technical report IEC/TR 80001-2-8 [31] to identify security controls that can be implemented in order to address the HIPAA safeguards.

IEC/TR 80001-2-8 provides a guidance "for the selection and implementation of management, operational, administrative and technical security controls to protect the confidentiality, integrity, availability and accountability of data and systems during development, operation and disposal." [31].

IEC/TR 80001-2-8 extends another security standard IEC/TR 80001-2-2 [32], which lists 19 security capabilities that are needed to ensure data security. IEC/TR 80001-2-8 extends IEC/TR 80001-2-2 by outlining a list of security controls for each of the security capabilities. The capabilities are broad and range from technical capabilities, e.g., "Automatic logoff" [31,32], to those that can be traced to documentation or policies, e.g., "Security guides" [31,32].

The mapping of security controls to security capabilities in IEC/TR 80001-2-8 is based on an extensive review of security controls outlined in the following security related standards and guidelines: NIST SP 800-53 (Revision 4) [33], ISO/IEC 15408-2:2008 [34], ISO/IEC 15408-3:2008 [35], IEC 62443-3-3:2013 [36] and ISO IEC 27002:2013 [37].

IEC/TR 80001-2-8 can be used to support the decision-making process when selecting security controls in order to achieve security capabilities for medical device software. The standard aims to help companies with identifying and implementing appropriate security controls to help "protect the confidentiality, integrity, availability and accountability of data and systems during development, operation and disposal." [31].

V. Current Status of the Case Company: Lessons and Challenges

First, once BBT became a MD software supplier they noticed the significant attention within the MD field. MD software manufacturers started to get in contact and invite tenders for various projects. In fact, to date they have worked on a number of MD software development projects for different types and sizes of manufacturers. Therefore, the overhead required to implement the necessary standards was starting to pay dividends. However, now that the opportunities clearly are out there it is noticeable that BBT now want to move to the next phase of their MD software development journey and not only develop software in line with the MD standards but their ambition is now to increase the efficiency of their MD software development. Therefore, they wish to improve their software development processes even further and implement more regulatory standards in relation to security etc. The key driver to take a step further is that BBT now are undertaking challenging projects and are developing MD software for multinational MD companies they have much more to achieve in their journey. BBT have agreed to work with researchers from the RSRC to introduce MD software development best practices that will increase the efficiency of their MD software development.

Second, BBT realized the increased attention to data security by the regulatory bodies and their USA clients. The increased awareness and value placed on data security by their clients means that implementing strong data protection mechanisms is now a competitive advantage. In addition, if appropriate security controls are not implemented as outlined by HIPAA, it can result in penalties [30, 38]. This can be detrimental to their business and stifle their growth and profits.

A. *Challenges for such large projects*

However, as with every new project there are associated challenges and this is increased when embarking upon a *fixed price* project, therefore if the project is delayed or runs into some other difficulties, BTT is liable in relation to the budget. Another challenge is the *tight timeframe* where *strict milestones* have to be achieved in addition to the achievement of appropriate documentation to satisfy *regulatory deliverables*. Additionally, BBT would also like to excel in being able to *facilitate change* during the lifecycle of the project as this is something that is challenging in traditional MD software development. A very positive aspect of BBT's current approach is that they engage in regular interaction with their customers. Therefore, receiving feedback and making sure that the right MD software is developed from the very start of the development.

B. *What is the current status of BBT development process lifecycle?*

Currently BBT is developing software in a plan driven way through using the V-model [19]. When following a V-model the testing is planned in parallel with the corresponding development phase and the planning for verification and validation of the product is emphasized from the very beginning. Even though V-model has been used by BBT successfully and it has been proven to be the best fit when developing MD software in compliance with the regulations [20]. However, in order to improve the efficiency of their software development new software practices should be explored that have proven successful in the development of safety-critical software in association with researchers from the RSRC.

Before introducing a new lifecycle it is crucial to perform an assessment in order to establish how the current software development process should be improved/changed.

VI. Assessment Process and Results

The following subsections will describe the high-level assessment process completed in BBT in 2016 and those in progress in 2017.

A) Software Development Process Assessment

The assessment carried out in 2016 was focused on the software development processes.

The Software development process assessment was performed at BBT before deciding what new practices would be most suitable for BBT. We met up with the CEO of the company, project manager/developer (who had has experience of agile software development), and a developer who specialized in Android software development. The meeting was also attended by the R&D manager/Systems Risk engineer and the QMS manager. The assessment was based on previously scripted open-ended questions that related to many different areas of the company as well as the software development process.

Results for the Software Development Process Assessment:

a) Currently BBT have several standards in place, such as IEC 62304, ISO 13485, ISO 9001 and ISO 14971. In their software development process they make use of various tools in areas such as project management, testing and integration. One of their main drivers for adopting new best practice software development methods is to streamline even further their already succcessful practices for interacting with customers. BBT view this as being key to delivering safe regulatory compliant software that fully meets the customer requirements and works within the intended environment, thereby decreasing the chances of expensive rework, particularly on fixed price projects.

b) Additionally, they wish to develop metrics such as problem tracking, code coverage, defects found, defects closed etc.

c) In the past BBT was open to changes and customers able to introduce them whenever they wanted without consequences to the overall budget, time. However, today the process has become more structured. BBT now ensures that a formal change document is in place specifying what happens if a change occurs within a previously signed project.

d) BBT at the moment is not making use of any principle software design techniques however, they plan to introduce architecture diagrams and design patterns.

e) BBT previously have developed software in a plan driven manner and lately they have decided to integrate some agile practices into their development process..

f) At the moment almost 80% of a testing is automated and 20% is done manually. If the percentage of manual testing could be decreased further – the overall development process could be faster. Automation of tests can prove challenging when components such as Bluetooth or Wifi are involved.

g) One of the team members mentioned that due to the new lifecycle approach where agile practices are introduced, there could be a challenges regarding integrating the QMS with the development process and achieving the necessary regulatory documentations.

h) At present their current process incorporates only two agile practices, they are: short iterations (every 2 weeks) and continuous integration.

i) BBT is also planning to provide their team with the training needed in order to work in an environment where MD software is developed in an agile way. The team will be provided with training in regards to MD software, agile practices and mobile app development.

j) Some team members will be provided with support to change towards adopting a more agile software development process.

B) Data Protection and Regulation Assessment

This section describes the high-level assessment process currently in progress at BBT in 2017.

BBT currently implements security controls to ensure data security. However, they need to ensure that their controls are inline with the safeguards outlined by HIPAA in order to continue working with their clients in the US. Strengthening data security will have the added advantage of enhancing their competitive advantage, avoid any regulatory risks in the future, and improve longevity within the MD software market. We are first assessing how well BBT's security controls for their software align with HIPAA safeguards. The focus is on the HIPAA safeguards outlined in the Security Rule. We are using a web-based tool to capture whether a particular HIPAA safeguard is either "fully implemented", or "alternative/partial security measure implemented to achieve the same purpose" or "not implemented and no alternative implemented". For each response further explanation will be captured to get details of what security control is implemented and how it isimplemented, or alternatively provide an explanation as to why no security control is implemented for a particular safeguard. An example is shown in Figure 2. The HIPAA safeguard shown in the example in Figure 2, "Access Control" is from the HIPAA Security Rule, under the Technical Safeguards.

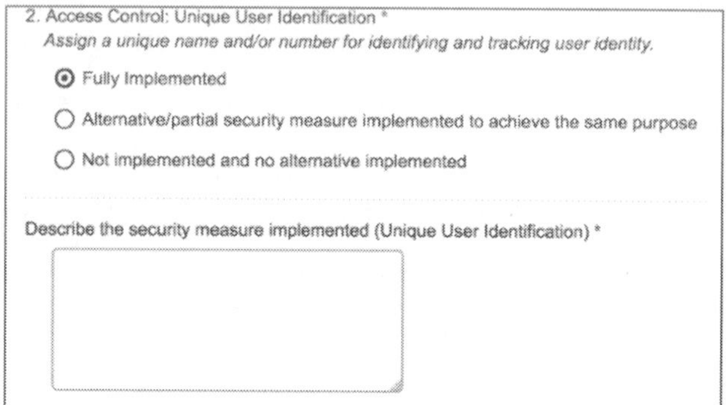

Figure 2. Assessment against HIPAA Safeguards

The next step will be data analysis and workshop with the developers at BBT to discuss gaps between the HIPAA safeguards and the security controls currently implemented. We will then use IEC/TR 80001-2-8 as a guide for selecting appropriate security controls for BBT to implement in order to address any HIPAA safeguards that are not well addressed. Relevant security risks will be taken into consideration during the process.

VII. RECOMMENDATIONS

A) Agility of Software Development Processes

Our advice to BBT is to integrate more agile practices into their current MD software development so that the software is developed efficiently in regular iterations and can be presented to the customer on a regular basis and facilitate change. Based upon a mini-literature review performed, the following agile practices have been cited as being used to develop software successfully for safety critical/medical domains:

a) Acceptance test-driven development (ATDD) [21].

b) Automated Tests/Automated unit testing [22].

c) Code Reviews / Peer Reviews [23].

d) Coding Standards [21], [24].

e) Continuous integration (CI) [21], [24], [25].

f) Open Workspace [21], [26].

g) Scrum [27].

h) Test-driven development (TDD) [21], [28]

B) Data Security: Adhering to HIPAA During Software Development

The safeguards for EPHI/PHI that can be traced to development work are outlined within the Security Rule [39]. They are outlined under the Technical and Physical Safeguards parts of the Security Rule [39].

The Security Rule has one other part, Administrative Safeguards, but this contains management processes, policies and planning, e.g., risk management and analysis procedures. The Safeguards in the Technical and Physical parts of the Security Rule are either labeled as ``required'' or ``addressable'' [29, 39]. Those labeled as ``required'' must be implemented. They must be included within the requirement specification document just to make sure that they are considered during design, implementation and the verification and validation phases of the software. The safeguards labeled as ``addressable'' can be implemented depending on how reasonable and appropriate they are given a particular context, e.g., the software and how it can be used. Alternatively, a different safeguard can be implemented if there are one or more safeguards that would achieve the same or better level of security, and are reasonable and appropriate. Figure 3 and Figure 4 show the steps that we propose for addressing the safeguards in the Technical and Physical Safeguards part of the Security Rule.

```
FOR each "addressable" safeguard
{
   IF: ("addressable" safeguard is reasonable and appropriate)
   {
      Then: Implement "addressable" safeguard
   Else IF: (one ore more alternative safeguard(s) are reasonable and ap-
propriate)
         Then: Implement one ore more alternative safeguards.
   Document process and reasons for selecting safeguard(s).
   Document how safeguard(s) is implemented.
   }
}
```

Figure 4. Implementing Addressable Safeguards

It is important to note that each of the security capabilities in the international standard IEC/TR 80001-2-8 comes with many security controls. But it is not practical or feasible to implement all of the security controls in a software product. This is because some of the security controls many not be relevant given the context or type of software security concerns for the software system. This is why, as suggested by the standard IEC/TR 80001-2-8, the selection of the controls should be based on risk or threat identification and assessment, and most importantly, both patient safety and protection of health information. It is also

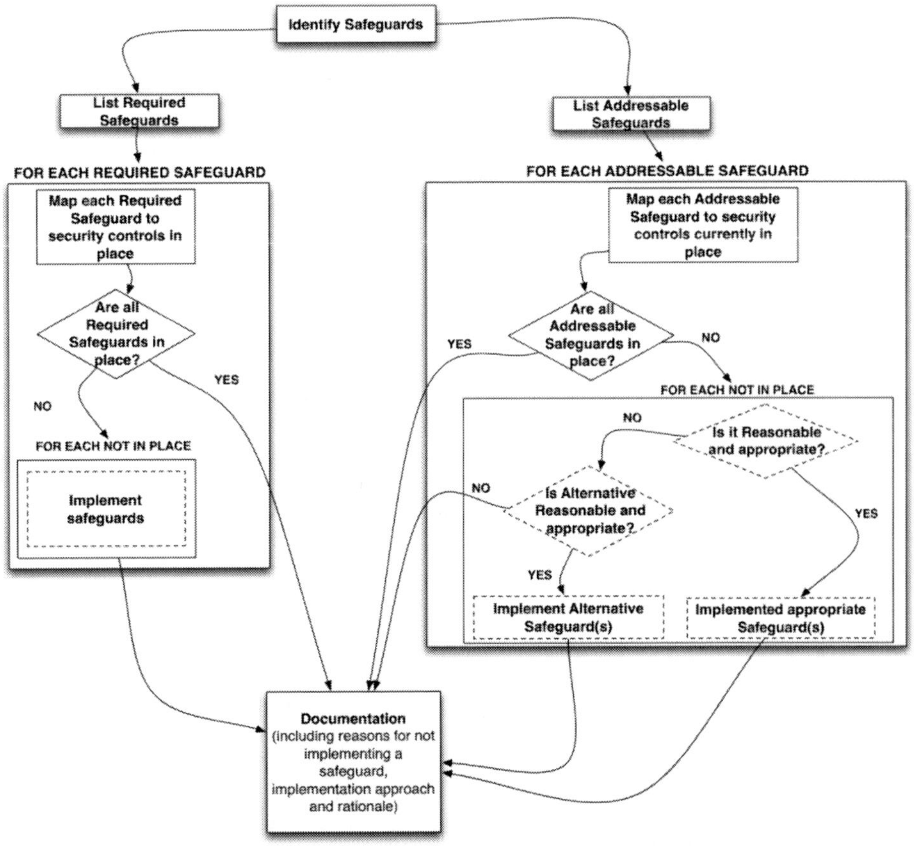

Figure 3. Proposed Approach for Addressing Security Controls for Technical and Physical Safeguards

worth noting that a similar consideration when implementing security controls is also advocated in the HIPAA Security Rule.

Understanding security threats and security weaknesses can help make an informed judgement in relation to the appropriate security controls to implement. Hence, input from a well detailed threat modeling approach will be very useful during the selection process. An overview of threat modeling is provided by The OWASP Foundation [40].

After implementation, maintenance of the safeguards, which should include improvements, should be performed as necessary in order to ensure continued protection of EPHI/PHI. This is particularly important for evolving software. The addition of new features or implementation of defect-fixes, which is part and parcel of software evolution, may affect the efficacy of the implemented safeguards. Therefore, continuous maintenance is necessary to ensure the implemented safeguards keep securing EPHI/PHI.

Our advice to BBT is to assign employees that continuously check compliance with HIPAA regulatory requirements, as well as on identifying and assessing relevant security threats and vulnerabilities. The employees responsibility will be to ensure that appropriate security controls are implemented, not only to comply with data security regulations, but also that relevant security concerns are addressed. The employees should be involved early within the development lifecycle, ideally from requirements elicitation, and continue throughout the evolution of the product. The identification and assessment of threats and vulnerabilities, which is advocated in the HIPAA Security Rule, can be done by following a threat modeling approach. As an example of a threat modeling approach we show in Figure 5 the one proposed by Oladimeji et al. [41]. More details of their well detailed approach can be found in their paper.

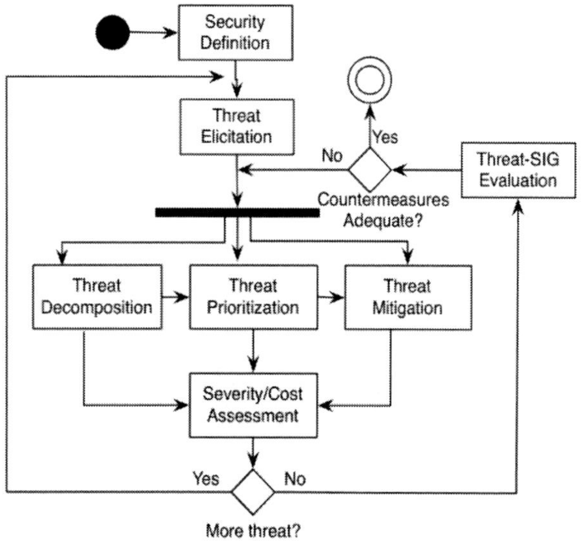

Figure 5. Threat modeling approach proposed by Oladimeji et al. [41]

The advantage for BBT, or any other company, of having dedicated employees that continuously assess compliance with data security regulations and assess threats and vulnerabilities is that it ensures that there is a proactive rather than a reactive process to addressing security concerns. This significantly reduces the likelihood of costly rework and implementing security patches late within the development lifecycle.

VIII. Conclusion and Future Work

This paper describes a case study of a journey taken by an Irish software development company, moving from developing automotive software to developing MD software. We described how through adopting and implementing MD standards they now have become a MD software supplier. Since becoming a MD software supplier many new opportunities have become available. However, BBT now wish to further improve their software development processes in order to become more efficient and to be able to satisfy new challenges that could rise from undertaking new multinational MD manufacturer's projects. They also need to take steps to ensure that their approach to ensuring data security is in line with regulatory requirements. Taking a software engineering approach, the authors of this paper provide a list of agile practices that have been cited to be well suitable for the safety critical/medical domain. The authors have also outlined an engineering approach to help with identifying and implementing appropriate security controls when developing software. The approach will help the company to develop software that is compliant with data security regulations.

In the future, we plan to investigate agile practices that are applicable for the MD software industry in greater detail by performing an extensive literature review and industry survey. Further, we will work with BBT to integrate the most applicable agile practices into their current software development lifecycle. We also plan to assist BBT with addressing data security concerns for the software that they develop. In addition, we will guide them through the process of ensuring that their software complies with HIPAA data regulations, and appropriate security controls are put in place. This is the first time that we are using a web-based tool for the assessment of how well the case company addresses HIPAA safeguards. We will use lessons learned from the process to refine and improve the tool.

Acknowledgment

This research is supported by the Science Foundation Ireland Research Centres Programme, through Lero - the Irish Software Research Centre (http://www.lero.ie) grant 10/CE/I1855 & 13/RC/2094.

References

[1] K. Trektere, F. McCaffery, G. Coady, and M. Gubellini, "Case Study: Becoming a Medical Device Software Supplier," *The Second International Conference on Fundamentals and Advances in Software Systems Integration*, (FASSI), 2016, pp. 24-29.

[2] J. Cunningham, B. Dolan, D. Kelly, and C. Young, "Medical Device Sectoral Overview," *Galw. City Cty. Econ. Ind. Baseline Study*, 2015.

[3] C. Abraham, E. Nishihara, and M. Akiyama, "Transforming healthcare with information technology in Japan: A review of policy, people, and progress," *Int. J. Med. Inform.*, vol. 80, no. 3, pp. 157–170, 2011.

[4] S. Hanna, R. Rolf, A. Molina-Markham, P. Poosankam, K. Fu, and D. Song, "Take two software updates and see me in the morning: The Case for Software Security Evaluations of Medical Devices," in *The 2nd USENIX conference on Health security and privacy*, 2011.

[5] S. R. Rakitin, "Coping with Defective Software in Medical Devices," pp. 40–45, 2006.

[6] EC, "Directive 2007/47/EC of the European Parliament and of the Council concerning medical devices," *Official Journal of the European Union*. Official Journal of the European Union, Brussels, Belgium, p. 35, 2007.

[7] F. McCaffery, J. Burton, A. Dorling, and V. Casey, "Software Process Improvement in the Medical Device Industry," in *Software Engineering Encyclopaedia*, P. Laplante, Ed. New York: Francis Taylor Group, 2010, pp. 528 – 540.

[8] C.-J. Hsiao, E. Hing, T. C. Socey, and B. Cai, "NCHS Health E-Stat Electronic Medical Record/Electronic Health Record Systems of Office-based Physicians: United States, 2009 and Preliminary 2010 State Estimates," *Natl. Cent. Heal. Stat.*, vol. 2009, no. December, pp. 6, 2010.

[9] G. Hillestad, R., Bigelow, J., Bower, A. and F. Girosi, "Can Electronic Medical Record Systems Transform Health Care? Potential Health Benefits, Savings, And Costs," *Health Aff.*, vol. 24, no. January, pp. 1103–17, 2016.

[10] "Mobile to Play a Significant Role in Healthcare as GSMA Research Predicts mHealth Market to be Worth US$23 billion by 2017," 2012. [Online]. Available: http://www.gsma.com/newsroom/press-release/mobile-to-play-a-significant-role-in-healthcare-as-gsma-research-predicts-mhealth-market-to-be-worth-us23-billion-by-2017/. [Accessed: 26-Mar-2016].

[11] BSI, "Medical device software- Software life-cycle processes, 62304:2006," *Bs En 62304:2006*, vol. 3, 2006.

[12] ISO, "ISO 13485: Medical Devices - Quality Management Systems - Requirements for Regulatory Purposes." Geneva, Switzerland, p. 57, 2003.

[13] ISO, "ISO 14971 - Medical Devices - Application of Risk Management to Medical Devices." Geneva, Switzerland, p. 82, 2009.

[14] F. McCaffery, M. Lepmets, and P. Clarke, "Medical Device Software as a Subsystem of an Overall Medical Device," in *Proceedings of The First International Conference on Fundamentals and Advances in Software Systems Integration Medical*, 2015, pp. 17–22.

[15] IEC, "IEC TR 80002-1 - Medical Device Software - Part 1: Guidance on the Application of ISO 14971 to Medical Device Software." Geneva, Switzerland, p. 58, 2009.

[16] FDA, "Guidance for Industry - FDA Reviewers and Compliance on Off-The-Shelf Software Use in Medical Devices." USA, p. 26, 1999.

[17] FDA, "Guidance for the Content of Premarket Submissions for Software Contained in Medical Devices." USA, p. 23, 2005.

[18] FDA, "General Principles of Software Validation ; Final Guidance for Industry and FDA Staff." USA, p. 47, 2002.

[19] K. Forsberg and H. Mooz, "The Relationship of System Engineering to the Project Cycle," *12th INTERNET World Congr. Proj. Manag.*, no. June 1994, p. 12, 1994.

[20] F. McCaffery, D. McFall, P. Donnelly, F. G. Wilkie, and R. Sterritt, "A Software Process Improvement Lifecycle Framework for the Medical Device Industry," in *IEEE International Conference and Workshops on the Engineering of Computer-Based Systems (ECBS'05) Proceedings*, 2005, p. 8.

[21] S. Datta, P. Sarkar, S. Das, S. Sreshtha, P. Lade, and S. Majumder, "How many eyeballs does a bug need? An empirical validation of linus' law," in *Lecture Notes in Business Information Processing*, 2014, vol. 179 LNBIP, pp. 242–250.

[22] J. Grenning, "Launching extreme programming at a process-intensive company," *IEEE Softw.*, vol. 18, no. 6, pp. 27–33, 2001.

[23] M. Bernhart, A. Mauczka, and T. Grechenig, "Adopting code reviews for agile software development," in *Proceedings - 2010 Agile Conference, AGILE 2010*, 2010, pp. 44–47.

[24] K. Beck, *Extreme Programming Explained*. 1999.

[25] P. Dahlem, Marc and Diebold, "Agile Practices in Practice - A Mapping Study Agile Practices in Practice - A Mapping Study -," in *18th International Conference on Evaluation and Assessment in Software Engineering*, 2015, no. MAY 2014.

[26] K. Beck, "Embracing change with extreme programming," *Computer (Long. Beach. Calif).*, vol. 32, no. 10, pp. 70–77, 1999.

[27] P. Abrahamsson, O. Salo, J. Ronkainen, and J. Warsta, *Agile software development methods Review and analysis*. ESPOO: VTT Publications 478, 2002.

[28] J. Rasmusson, *The Agile Samurai, How Agile Masers Deliver Great Software*. Texas, 2010.

[29] The Health Insurance Portability and Accountability Act

(HIPAA) of 1996. Pub. L. 104-191. Stat. 1936.

[30] Brief OCR Privacy. "Summary of the HIPAA Privacy Rule." Washington, DC, United States Department of Health and Human Services, 2005.

[31] IEC 80001-2-8 - Application of risk management for it networks incorporating medical devices – part 2-8: Application guidance – guidance on standards for establishing the security capabilities identified in IEC 80001-2-2. International Electrotechnical Committee, pages 43.

[32] IEC 80001-2-2 - Application of risk management for it-networks incorporating medical devices-guidance for the disclosure and communication of medical device security needs, risks and control. International Electrotechnical Committee, pages 48.

[33] NIST, SP. "800-53." Recommended Security Controls for Federal Information Systems: 800-53, 2013.

[34] ISO/IEC 15408-2:2008 - Information technology -- Security techniques -- Evaluation criteria for IT security -- Part 2: Security functional components. *International Organization for Standardization / International Electrotechnical Committee*, pages 218.

[35] ISO/IEC 15408-3:2008 - Information technology -- Security techniques -- Evaluation criteria for IT security -- Part 3: Security assurance components. *International Organization for Standardization / International Electrotechnical Committee*, pages 174.

[36] IEC 62443-3-3:2013 - Industrial communication networks - Network and system security - Part 3-3: System security requirements and security levels. International Electrotechnical Committee, pages 80.

[37] ISO/IEC 27002:2013 - Information technology -- Security techniques -- Code of practice for information security controls. International Organization for Standardization / International Electrotechnical Committee, pages 80.

[38] S. Hoffman and A. Podgurski, "Securing the HIPAA Security Rule. Journal of Internet Law", Spring 2007; Case Legal Studies Research Paper No. 06-26.

[39] TrueVault, HIPAA compliance developer guide. TrueVault, 2014. Retrieved from https://github.com/truevault/hipaa-compliance-developers-guide. Accessed on 27th Februrary 2017.

[40] OWASP Foundation "Threat Risk Modeling", Retrieved from https://www.owasp.org/index.php/Threat_Risk_Modeling. Accessed on 27th Februrary 2017.

[41] E. A. Oladimeji, S. Supakkul, and L. Chung. "Security threat modeling and analysis: A goal-oriented approach." Proc. of the 10th IASTED International Conference on Software Engineering and Applications (SEA 2006). 2006.

The Use of the Forest Status Quality Indicator in Planning Policies for Biodiversity Conservation

Ten Case Studies on Forest Plantations in North Italy

Silvia Assini
Dept. of Earth and Environment Sciences
University of Pavia
Pavia, Italy
Email: silviapaola.assini@unipv.it

Maria Grazia Albanesi
Dept. of Electrical, Computer and Biomedical Engineering
University of Pavia
Pavia, Italy
Email: mariagrazia.albanesi@unipv.it

***Abstract*—This paper describes a new application of the previous defined Forest Status Quality Indicator in planning policies of forest conservation. Starting from previous analysis and conclusions about the importance of the forest plantations for ecological restoration, new case studies confirm the possibility of using this indicator to efficiently guide policy makers and planners to choose the best position and extensions of forest types/patches for restoration and connectivity. By considering the present situation of the forest quality, we propose a method to compute a predicted value of the indicator by considering, in addition, the contribution of plantations, once they will be naturalized in the territory under consideration. The new, predicted value of the Forest Status Quality Indicator is a function of the ecological components, which are at the basis of the naturalistic approach of its definition. The results are shown on ten important case studies of plantations in the West Po Plain (North Italy). The method of prediction can be used to measure the impact of plantations on the quality of the forests and to define under which hypothesis they can play an important role in biodiversity conservation.**

Keywords-biodiversity; environmental indicator; forest status quality; Geographic Information Systems; conservation policy.

I. Introduction

This paper is based on the authors' work [1] and reports the extension of their current research activity. Compared to [1], nine more case studies are added; moreover, the present paper provides the list of all the forest types found in the studied locations, which was not possible in the previous work. Finally, the theory about the ecological indicator, called Forest Status Quality (FSQ), already reported in [1], has been validated by showing a new application to the definition of policies for forest management and conservation.

The previous target, i.e., to propose and measure indicators related to some specific aspects of biodiversity, within a given territory, has been extended to show how an indicator can be used to assess the efficacy of a policy of forest conservation.

The naturalistic indicator FSQ has been introduced for the first time in literature with an investigation about its relationship with another ecological indicator, the land use Anthropentropy Factor [2][3]. However, the preliminary results were very promising in giving a realistic assessment of the situation of forests quality in a given territory.

For this reason, the research has been carried further along this direction. Here, we recall from the previous contribution [1] some important aspects of FSQ indicator, in order to appreciate the improvement of the ongoing researches.

In literature, the most common approach to define biodiversity indicators is to use separately the following primary attributes of biodiversity: (1) species/composition, (2) structure, and (3) function [4], or landscape metrics [5].

The main innovative feature of FSQ, with respect to literature, is the joined evaluation of species composition and structure and landscape metric. In fact, the biodiversity components are: the stratification, the percentage of alien species and the percentage of protected species. Stratification can be easily depicted in four layers: tree, high and low shrub, and herb layers (see Figure 1). Figure 2 shows some typical alien species in the territory under investigation. On the other hand, often protected species (see Figure 3) correspond to true forest species, such as *Anemone nemorosa*, *Campanula trachelium*, *Carex elongata*, *Convallaria majalis*, *Listera ovata*, *Neottia nidus-avis*, and *Primula vulgaris*. The three components characterize the different forest patches of the territory under investigation of the ten case studies.

Furthermore, the FSQ indicator takes into account also a landscape feature, i.e., the size of the forest patches. Consequently, only patches greater than 10.000 square meters are considered. In fact, patches smaller than 1 ha generally show low species richness [6] and a scarce floristic quality due to the edge effect which can increase the abundance of weedy and alien species [7]-[9]. The richness and floristic quality (due to true forest species) of the forest patches can be influenced not only by their size, but also by their shape. However, a correlation between the shape and the species richness of forest patches can be found when the patch size is sufficiently high. In fact, Dzwonko and Loster [10][11] found a negative correlation between the shape index of Patton [12] and the number of shrubs and forest species. In that case, they worked with a restricted dataset of only 27 forests, with varying patch size from 0.03 to 1.6 ha. With such small patch sizes, it is possible that the entire patch was subject to the edge effects [13]. Honnay et al. [13] analyzed 234 forest patches varying

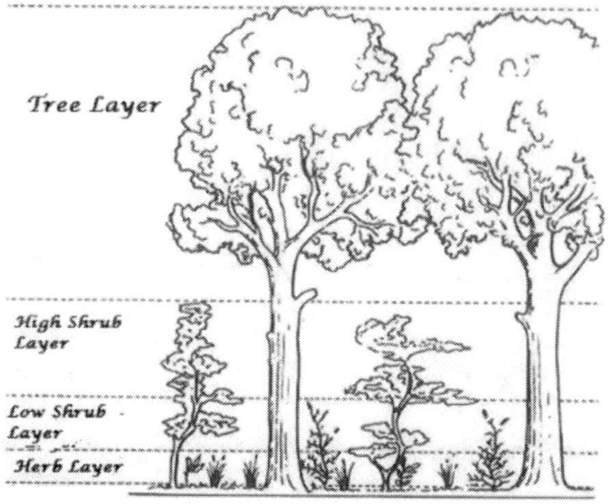

Figure 1. The layers of the stratification in a forest: tree, shrub (high and low), and herb layers.

Figure 2. Examples of some alien species in the territories under investigation: *Solidago gigantea*, *Robinia pseudoacaica*, and *Amorpha fruticosa*.

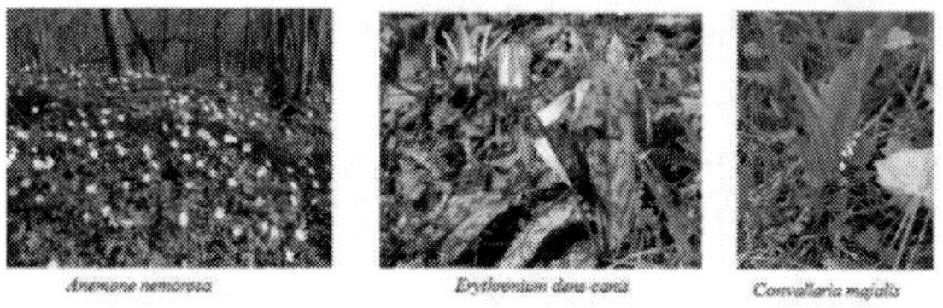

Figure 3. Examples of some protected species in the territories under investigation: *Anemone nemorosa, Erythronium dens-canis,* and *Convallaria majalis.*

in size between 0.5 and 5216 ha and found a correlation between the Patton shape index and the number of species in edges and clearings, the number of woody species and lianas and, as a consequence, with the total number of forest plant species. For these reasons, in a second improved definition of the FSQ indicator [1], also the shape of forest patches has been taken into account.

Patches with linear shapes are excluded, applying an ad hoc morphological operator on the image data set provided by the Geographic Information Systems (GIS) of the project (as explained in Section II.) Such linear patches, together with those smaller than 10.000 square meters, that can be considered as punctual patches, do not represent core areas hosting complex forest ecosystems (as the wide patches of compact shape where the edge effect is reduced) important for woody and herbaceous true forest species. However, they can represent critical elements to connect core areas and support the structure of a local/regional ecological network, and the evaluation of their quality should be better considered from the connectivity perspective. Therefore, only forest patches with similar geometric characteristics (wide size and compact shape) were considered, with the aim to evaluate more homogenous elements and to have a more realistic picture of the forest quality, which is useful for conservation purposes.

Besides embedding the geometric characteristics in the computation of the FSQ indicator, a second aspect has been considered [1]: the evaluation of the importance of forest plantations. This idea is at the basis of the new research activity here reported. In fact, in the first elaboration of the FSQ indicator [2][3], only natural forests were considered, while plantations were excluded from the evaluation. However, the Region Lombardy, in 2002, started and financed an important project aimed to create ten new plain forests, each of them with a size of about 40 ha and planted with native trees and shrubs [14][15].

Actually, such forests have not still developed a typical structure of mature wood and a typical nemoral herb layer, which will require at least 20-30 years. Anyway, because of their importance for the restoration of the Lombardy plain and conservation purposes, the initial idea of considering also such new forests give interesting preliminary results on a first case study in the Municipality of Travacò Siccomario (Province of Pavia).

After these preliminary results [1], in this paper we extend the analysis on other nine great forest plantations of the Region Lombardy, with an innovative approach to predict the efficacy of these plantations in their contribution to FSQ. This will be further explained in Section III. First, a brief description of preliminary researches on the theory about FSQ is given in Section II. Conclusion and future work end the paper in Section IV.

II. Preliminary Research activities

In this section, a brief summary of the theory of FSQ indicator is given. Moreover, preliminary results are reported with more details in [1], in order to appreciate how the research has been evolving to make improvements and new contributions with respect to the state of the art.

A. *Definition of the Forest Status Quality Indicator*

The FSQ expresses the forest quality status as the value of its ecological components, with particularly reference to the biodiversity conservation. A set of sub-regions occupied by natural forest F_i (i = 1, 2, n) was defined. Each of F_i may have one or more occurrences, denoted by the index k, in the territory (k = 1, 2, max(i)). The number of occurrences may vary from a minimum of 1 to a maximum, which depends on the forest type (max(i)). Each k-th occurrence is characterized by: (a) an area A^k_i, expressed in square meters, for i = 1, 2,...n and k = 1, 2, ... max(i), and (b) a type of T_i, derived from the GIS ERSAF Database "Map of the Forest Types of Lombardy" [16]. For each forest T_i, we found the correspondence with one or more phytosociological tables [17]. For each forest type T_i, we defined a set of the following indicator components (s_i, a_i, p_i): the stratification (number of layers) of a forest type $_i$ (s_i), the percentage frequency of alien species (a_i) in the corresponding phytosociological table/s, and the percentage frequency of protected species (p_i) (according to the Lombardy regional law, L.R. 10/2008) in the corresponding phytosociological table/s. The three components can assume only discrete values, from 0 to 3, according to an if – then – else algorithm described in [2] [3]. After determining the values of the set of components for stratification, alien and protected species, for each forest T_i, it is possible to compute the FSQ Indicator of a municipality of area S as

$$FSQ = \Sigma_i \ \Sigma_k (s_i + a_i + p_i)*A^k_i/S \qquad (1)$$

The FSQ definition is the weighted values of the components, where the weights are the ratios between the areas of the forest patches and the area of the territory under investigation. By using the primitives of the open source QGIS software [18], the values of A^k_i in the territory under investigation have been computed, in order to estimate the value of the FSQ indicator, according to (1), for all the municipality of our case studies. Obviously, according to (1), FSQ indicator is a real number in a limited range: the minimum value is 0, which corresponds to the dramatic situation where no forests are present in the territory with at least one occurrence of area greater than 10.000 square meters. The maximum possible value is 9, which is derived by considering the unrealistic situation of a municipality where no anthropic presence occurs, and the territory is entirely occupied ($\Sigma_t \ \Sigma_K \ A^k_i = S$) by forests of very high quality (set of components (s_i, a_i, p_i) = (3,3,3)).

TABLE I. THE METRIC ON THE *FSQ* INDICATOR FOR FOREST QUALITY.

Class of forest quality	Evaluation of Forest quality and policy	
	Intervals of FSQ	*Suggested policy*
1 Unsatisfactory	0 <= FSQ <= 0.9	Very low level forest quality. A high-impact policy of restoration and/or requalification of forest is mandatory.
2 Satisfactory but improvable	0.9 < FSQ <= 1.8	Sufficient forest quality but improvable. A policy for forest biodiversity preservation is preferable.
3 Good	1.8 < FSQ <= 3.6	Good forest quality, the first level of satisfactory situation.
4 Optimum	3.6 < FSQ <= 4.5	The optimun situation, with a high quality of forests.
5 Overbalanced	FSQ > 4.5	The overbalanced situation, forests overcoming other ecosystems. A policy for shrubland and grassland biodiversity preservation is preferable.

We have defined a set of ranges for the FSQ indicator, starting from an unsatisfactory forest quality, a satisfactory but improvable situation, a good, an optimum situation and overbalanced situation. In Table I, the metric for the FSQ indicator and the suggested policy actions are shown.

B. *The Erosion Operator*

The introduction of the factor "shape" of the forest patches in the computation of the FSQ slightly modify the original definition in (1), where only the areas of forest patches are considered and weighted by the naturalistic components (s_i, a_i, p_i). The only limitation introduced on the geographic data is still a quantitative one (the 10.000 square meters as the minimum accepted size of the forest patch to be evaluated). However, we decided to take into consideration a characteristic that refers to the shape of a patch, i.e., to reduce the edge effect, which, in turn, reduces the floristic quality. This can be done by applying the standard mathematical morphology operator of Erosion [19], with a structural element of a circle of radius of 50 meters. In fact, Erosion is a typical image processing operator that allows to "erode" a connected area, starting from its perimeter and proceeding inside, of an extent that corresponds to the shape of the structural element. In our case, the structural element is a circle of a given radius, in order to reduce the areas of the forests to their real inner shape, by excluding the areas near the boundary. If we use a circle as structural element, the shape of the forest will be remodeled in a symmetric way, all along the boundaries. The diameter of the structural element determines the minimum distance that a forest patch must have from its center to all the points of its boundaries, in order to be considered in the FSQ computation.

As a result of the Erosion, the forest patches with a linear shape (a thin stripe less than 100 meters of amplitude) disappear from the map, as it can be seen in the experimental results reported in Section II.E. All the other patches are reduced by the erosion, to minimize the "edge" effect. With the introduction of the Erosion operator, we applied to each area A^k_i the Erosion operator E[], thus leading to a new expression of FSQ, denoted by FSQ_e:

$$FSQ_e = \Sigma_i \ \Sigma_k (s_i + a_i + p_i)*E[A^k_i]/S \qquad (2)$$

The new operator FSQ_e takes into consideration both the quantitative (areas) and the qualitative (shapes) aspects of the forest patches.

C. *The hardware and software infrastructure*

In order to compute the values of FSQ in (1) and (2), we settled a software environment, capable of measuring the areas of the territories of the municipalities under investigation, the areas of the forest patches, and to implement the Erosion operator. The same hardware/software infrastructure has been used both for the previous results [1][2][3], and for the new research activity here presented.

The hardware resource is a standard Workstation with a 64 bit Intel® Core^{TM} I7-3630QM microprocessor at 2.4 GHz, with 12 GB RAM with a Full HD resolution monitor, for easily displaying details of the maps, and Windows 10 operating system.

The software equipment is based on QGIS software [18], 2.14 version, for visualizing, managing and processing geospatial information. In fact, as in the previous contribution [1][2][3], we have used a GIS map database (ERSAF database), free available from the Region Lombardy [16], in order to derive the useful description about the distribution of the forest in the territory. In Figure 4, the visual map for a portion of the Municipality of the preliminary case study [1] is shown. Different colors refer to different types of forests. Particularly, the compact pink area refers to the forest plantation. The other colors refer to natural forests.

The second GIS map we used is provided by ISTAT, the Italian National Institute of Statistics, for the administrative boundaries of the municipalities [20]. By using the QGIS primitives, it is possible to compute on this map the area of the municipality under investigation, i.e., the value of S in (1) and (2). By combining information of the two GIS maps [16][20], it is possible to define exactly which of the Forests Types of the ERASAF Database belongs to each Municipality under investigation.

The Erosion operator (see Section II.B) has been implemented by using the primitives of morphological operators in Matlab environment [21] for image processing. Finally, some *ad hoc* software has been developed for the generation of the quantitative results of Section III and their storing and management in a typical relational database, to facilitate access and subsequent use in future researches.

D. The Study Area

The study area in the preliminary research [1] is the Municipality of Travacò Siccomario. It is located in the Lombardy plain at an average altitude above sea level of 66 meters, and is mainly characterized by agricultural fields and urban areas. Natural forest areas are localized along watercourses and channels, often as linear elements dominated by mixed woody species as *Salix alba*, *Populus alba*, *Quercus robur*, *Ulmus minor* and, very frequently, the invasive *Robinia pseudacacia*. Wide forest patches are limited and dominated by the woody species above mentioned. As we consider only forest types that have at least one occurrence in the territory of area greater than 1 ha, only three types of natural forests survive these requirements: the *Salix alba* communities, the *Populus alba* communities, and *Robinia pseudoacacia* communities.

Moreover, in this study we want to include also plantations in the FSQ computation. In 2003-2004, one of the great plain forests of the Lombardy Region was realized in this municipality. This project was entitled "The Great Forest between the Two Rivers", because it is at the join of Ticino with Po River (see Figure 5). In particular, meso-xerophilous and meso-igrophilous forest patches were realized on a surface of about 41 ha, planting native trees (such *Populus nigra, Ulmus minor, Acer campestre, Malus sylvestris, Carpinus betulus, Quercus robur, Salix alba, Alnus glutinosa, Populus alba, Prunus padus*) and native shrubs (such *Crataegus monogyna, Corylus avellana, Prunus spinosa, Sambucus nigra, Cornus sanguinea, Frangula alnus, Viburnum opulus, Salix cinerea*). Considering the woody floristic composition, the new forest can be considered as the forest type Oak-Elm wood (also including the Black Alder variant) [22]. The value set of components for stratification, alien and protected species is 3,2,2 for this plantation. In Table II, all the forest types, of area greater than 1 ha, for the case study of Travacò Siccomario are shown.

TABLE II. FOREST TYPES FOR THE PRELIMINARY CASE STUDY [1].

Naturalistic components (s_i, a_i, p_i)	**Description of forest types and relative reference *syntaxa***
(1,1,0)	Willow wood of bank
(3,1,0)	White Poplar formation
(2,1,0)	Pure *Robinia pseudoacacia* wood
(3,2,2)	Plantation (Oak-Elm wood, also including the Black Alder variant)

Figure 4. The territory under investigation (Municipality of Travacò Siccomario) in the preliminary case study [1]: the blocked colored areas superimposed on the geographic map show the presence of the natural forests (light greeen and light blue) and the plantation (pink).

Figure 5. The "Great Forest between the Two Rivers" forest plantation project: an overview of the territory where river Ticino joins the river Po.

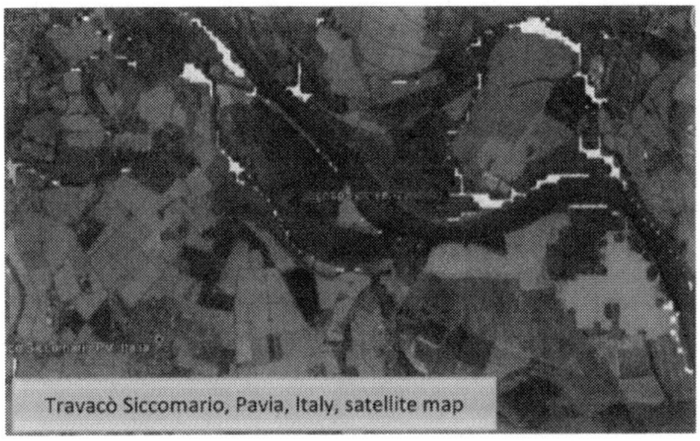

Figure 6. The territory under investigation after the Erosion operator applied to all the forest patches.

E. Preliminary Experimental Results

By applying the erosion operator to compute the FSQ_e, according to (2), we can compare the effect of the Erosion on each forest types of Table II. Clearly, the erosion has reduced the areas of the forest patch. However, the loss of the areas is not equal for all the four forest types. In fact, the loss is determined by the shapes of the patches, which are quite different in the territory under examination. The loss percentages are the following: 3.6% for the plantation, 35.2% for the *Salix alba* communities, 51.3% for the *Populus alba* communities, and 61.2% for the *Robinia pseudoacacia* communities. It is interesting to note that the plantation is the forest type with the lowest loss; this means that in the project the plantation patch has a shape, which is very close to an ideal core area. Moreover, the *Robinia pseudoacacia* communities have the highest loss, and it is a positive aspect, because of the very low floristic quality, due to the dominance of alien species in linear and fragmented patches. In Figure 6, the map of the territory after the erosion is shown. The reduction of the colored areas of the forests types is evident, by comparing Figure 6 to Figure 4. In Figure 7, the areas of the forests (Y axis) before and after the erosion are plot, for each forest type of Table II.

III. Planning Conservation Policies using the FSQ Indicator

In the preliminary study [1], we have evaluated the naturalistic components of the plantation in the municipality of Travacò Siccomario. This initial idea can be further investigated by directly explaining the contribution of plantations in the definition of the FSQ indicator. This operation is computed starting from the original formula (1) of FSQ, without the erosion operator, because, according to the results reported in Section II, the Erosion operator is quite irrelevant on the plantations (Figure 7), due to their compact shapes.

Suppose that, in a given municipality under investigation, there is a plantation of area A_p, with its naturalistic components (s_p, a_p, p_p). For simplicity, we show the discussion in the case of a single plantation, but the theory can be easily extended to the case of more than one patches. By reconsidering formula (1), we can separate the contribution to the global value of FSQ in two distinct parts: the first is due to the natural forest patches in the territory, the second is due to the plantation only. A new definition of FSQ is possible, which considers also the plantation contribution. We denote this "new" FSQ with the symbols FSQ', and (1) becomes:

$$FSQ' = \Sigma_i \ \Sigma_k (s_i + a_i + p_i)*A^k_i/S + (s_p + a_p + p_p)*A_p/S \quad (3)$$

Usually, a plantation have not still developed a typical structure of mature wood; moreover its naturalistic component can be only estimated, in its actual value of the triplet (as in the first case study of Travacò Siccomario, where (s_p, a_p, p_p) has been set to (3,2,2)), but we have no assurance that these values will be the same after 20-30 years, when the plantation will become, presumably, a new mature and "natural" forest. For this reason, it is interesting to study the variation of FSQ' in (3), for each possible values of the triplet (s_p, a_p, p_p). This analyses can express the strength of the naturalistic components in rising (or lowering) the FSQ' indicator. By analyzing (3), we can observe that the sum of the three naturalistic components (s_p, a_p, p_p) is a linear, multiplicative coefficient of the second term. In order to reduce the complexity of the study (actually, in (3) there are four independent variables, i.e., the area of the plantation A_p and the three components (s_p, a_p, p_p), we can substitute in (3) one variable x defined as

$$x = s_p + a_p + p_p \quad (4)$$

where x can assume only discrete values, in the range [1,9]. By observing that in (3) the first term corresponds to the FSQ value, computed without plantations according to (1), by substituting the value of FSQ (1) and x (4) in (3), we can write

$$FSQ' = FSQ + x*A_p/S \quad (5)$$

where FSQ' is the new value of the FSQ indicator, which includes plantations too, FSQ is the original value, which excludes plantation, according to the initial hypothesis [2], [3], x is the independent variable and A_p is the area of the plantation. The new definition of FSQ' can be used to reach two different goals: (a) if the plantation is already present in the territory, by varying x, we can compute the minimum and maximum gain of the FSQ value; (b) if the plantation is not still present, but it is the object of a planned, future project of a policy for forest conservation, it is possible to study as the indicator can improve, from its actual value of FSQ to the new value, i.e., FSQ', and consequently we can also determine to what extent the class of forest quality of

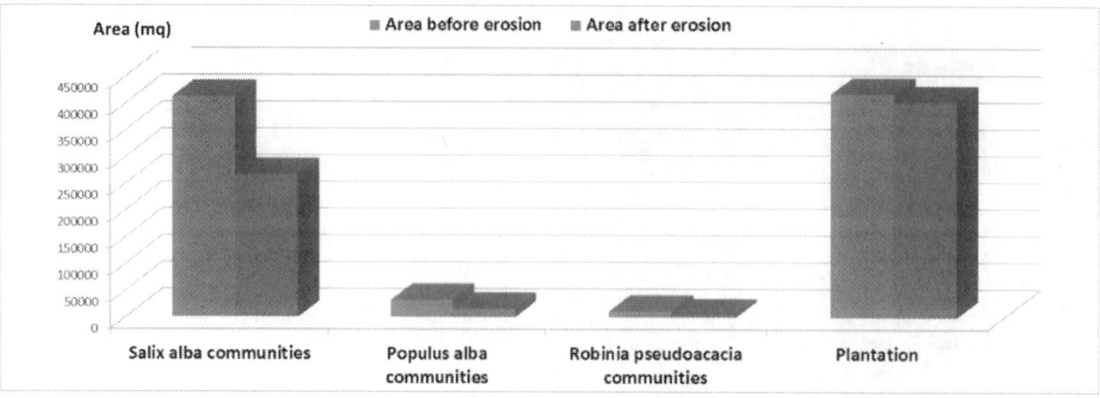

Figure 7. The areas (in square meters, on Y-axis) for each the forest types (on the X-axis), before and after the Erosion.

the territory may improve (according to Table I). The two cases will be analysed in details in Section III.A and Section III.B, respectively.

A. *Ten case studies of existing plantations*

We consider ten case studies, which are part of the project, "Ten Great Forests for the Plain", in Region Lombardy, in the North part of Italy (Figure 8). The project is the result of two years of intensive work toward a goal strongly shared by Regions, Provinces, Municipalities, ERSAF "Regional Agency for Services to Agriculture and Forests" and several other organizations and associations in the area. At present, eight forests have been funded, and seven have been completed. By careful analysing the ERSAF GIS database [16], we have identified further three great plantations in Region Lombardy. Therefore, we consider ten case studies, listed in Table III. Their positions in Region Lombardy are depicted in Figure 9. For completeness, in Table IV, a list of the natural forest types located in the territories of the ten case studies is provided, together with their reference *syntaxa* and the values of the triplets of naturalistic components. The *Type Lab* field in the Table is a label, which refers to the ERSAF database [16] used as input data source, and it is reported to help any comparative analysis and reference to the database. Moreover, in Table V, the membership of each forest type to each case study is reported.

By using the primitives of the open source QGIS software [18] on the GIS ERSAF data set [16], for each municipality of the ten case studies, we have computed the following values:

- the FSQ value (according to (1)) ;
- the areas of all the plantations (the term A_p in (5)).

At this point of the analysis, we can obtain a first important set of results: for each case study, it is possible to plot FSQ', according to (5), as a function of x (see(4)), i.e., as a function of the forest quality of the plantation, expressed in terms of its naturalistic components of stratification, alien and protected species.

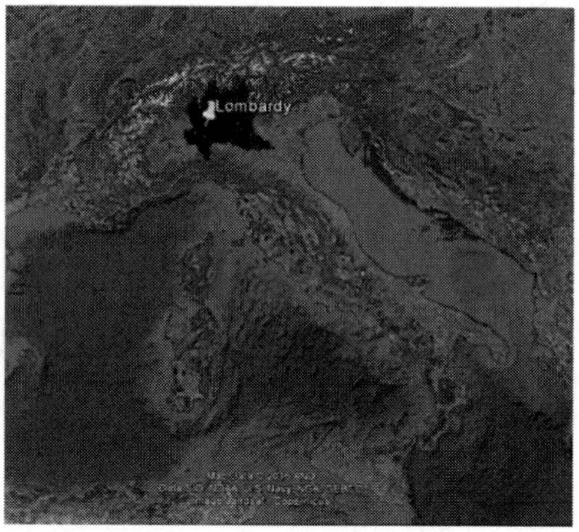

Figure 8. The geographic position of Region Lombardy in the North West part of Italy.

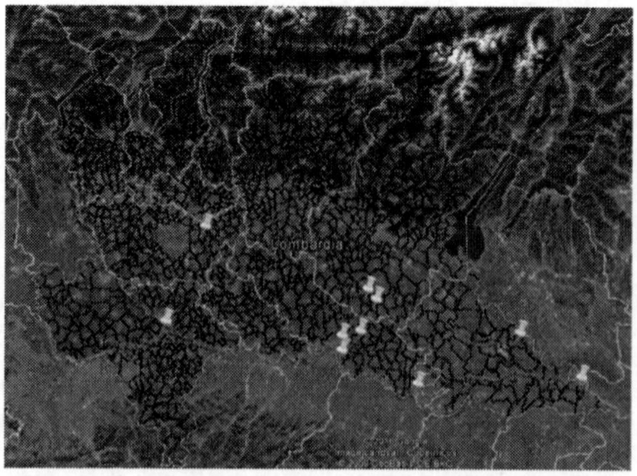

Figure 9. The ten case studies in the Region Lombardy.

TABLE III. THE TEN CASE STUDIES OF EXISTING PLANTATIONS: DENOMINATION AND GEOGRAPHICAL COORDINATES.

Case study	Name of the Municipality	Geographic Position	
		Latitude	Longitude
1	Travacò Siccomario	45° 9'7.58"N	9° 9'42.82"E
2	San Gervasio Bresciano	45°18'22.41"N	10° 8'54.31"E
3	Milzano	45°16'22.29"N	10°11'29.93"E
4	Cremona	45° 8'25.53"N	10° 1'55.86"E
5	Gadesco-Pieve Delmona	45° 9'38.92"N	10° 7'19.31"E
6	Gerre De' Caprioli	45° 5'26.18"N	10° 2'7.59"E
7	Casalmaggiore	44°59'17.91"N	10°25'10.63"E
8	Bigarello	45°10'55.20"N	10°54'44.63"E
9	Carbonara di Po	45° 2'16.78"N	11°13'26.36"E
10	Pioltello	45°29'22.90"N	9°19'34.25"E

TABLE IV. FOREST TYPES IN THE TEN CASE STUDIES: REFERENCE SYNTAXA AND NATURALISTIC COMPONENTS.

Type Lab[a]	Naturalistic components (s_i, a_i, p_i)	Description of forest types and relative reference *syntaxa*
1	3,2,3	Oak-Hornbeam wood of the lowlands *Syntaxa: Polygonato multiflori-Quercetum roboris* subass. *carpinetosum* and *anemonetosum* Sartori 1984; *Quercus robur, Carpinus betulus* and *Physospermum cornubiense* community; *Quercus robur, Carpinus betulus* and *Holcus mollis* community
14-15	3,2,2	Oak-Elm wood (also including the Black Alder variant) *Syntaxa: Polygonato multiflori-Quercetum roboris* subass. *ulmetosum* Sartori 1984
172	3,3,1	Black Alder wood of gulley *Syntaxa: Alnus glutinosa, Populus alba* and *Ulmus minor* community
173	2,3,2	Typical Black Alder wood *Syntaxa: Osmundo regalis-Alnetum glutinosae* Vanden Berghen 1971; *Carici elongatae-Alnetum glutinosae* W. Koch 1926 et R. Tx. 1931; *Carici acutiformis-Alnetum glutinosae* Scamoni 1935
177	1,1,0	Willow wood of bank *Syntaxa: Salix alba* community; *Salicetum albae* Issler 1926
183	3,1,0	White Poplar formation *Syntaxa: Populus alba* community
188	2,1,0	Pure *Robinia pseudoacacia* wood *Syntaxa: Robinia pseudoacacia* community
189	3,2,0	Mixed *Robinia pseudoacacia* wood *Syntaxa:Robinia pseudoacacia, Quercus robur* and *Ulmus minor* community

a. According to ERSAF database [16]

The fundamental question is: what is the improvement of the forest quality due to the presence of the plantation, expressed by the indicator FSQ', if compared to the situation where plantations are not present or considered (FSQ value)?

To find the answer, it is convenient to plot the absolute gain of FSQ' over FSQ, as a function of x. By considering (5), we can express the absolute gain as

$$FSQ'/FSQ = 1 + x * A_p/[S*FSQ] \quad (6)$$

In Figures 10 and 11, the absolute gain in (6), expressed as a percentage, is plot as a function of x, for the ten case studies. They are grouped in two sets: the first one (Figure 10) refers to the municipalities that show a modest increment of the FSQ' (*low increasing cases*), if compared to FSQ (less than 5%). The second set shows very interesting gains (*high increasing cases*), up to 90%. Obviously, Equation (6) is a straight line, where the slope is directly proportional to the area of the plantation (A_p) and inversely proportional to the area of the municipality. For this reason, case studies where the area of the plantation is considerably less that the area of the municipality show less values of gain.

The case study #8, Bigarello, suggests important considerations about policy of forest conservation. In fact, as it can be inferred from Table V, in the municipality of case study # 8 there are not natural forests of areas greater than 1 ha; therefore, the value of FSQ is zero. For this reason, this case study does not appear in Figure 10 and 11, because the absolute gain is numerically, equal to infinite (FSQ = 0 in (6)).

The plots of Figures 10 and 11 are very interesting because they show a potential very impressive gain in the Forest Status quality Indicator, if we choose carefully the area of the plantation and also the forest types involved, i.e., their naturalistic components, expressed by the cumulative variable x.

For this reason, we have analysed the projects of the plantations and, considering the planted species, we have defined the corresponding forest types (see Table VI) with their current value of the triplet (s_p, a_p, p_p) for each case study, as follows:

Case Study #1: $s_p = 3; a_p = 2; p_p = 2$
Case Study #2: $s_p = 3; a_p = 3; p_p = 3$
Case Study #3: $s_p = 3; a_p = 3; p_p = 3$
Case Study #4: $s_p = 3; a_p = 2; p_p = 2$
Case Study #5: $s_p = 3; a_p = 2; p_p = 2$
Case Study #6: $s_p = 3; a_p = 2; p_p = 2$
Case Study #7: $s_p = 3; a_p = 2; p_p = 2$
Case Study #8: $s_p = 3; a_p = 3; p_p = 3$
Case Study #9: $s_p = 3; a_p = 3; p_p = 3$
Case Study #10: $s_p = 3; a_p = 2; p_p = 3$ (7)

By substituting (4) and (7) in (6), we can obtain the absolute gain of the forest plantation, for each case study. The final results are summarized in Table VII. Some of them are noteworthy: for example case study # 8 goes from an initial dramatic situation (FSQ = 0, i. e., no forests) to a limited but interesting little improvement (FSQ' = 0.13).

Furthermore, in the case study #6, thanks to the new plantation, the municipality is able to improve its class of forest quality, from class 1 (FSQ = 0,014784877, thus < 0.9, See Table I), to class 2 (FSQ' = 0,925521). In the other cases, the increasing in the FSQ' is not sufficient to determine also an improving in the forest quality class. However, the absolute gain (expressed in percentage in Table VII) can be relevant, as in the case studies #2, #7, and #10 and, presumably, an enlargement of the plantation will be sufficient to improve the forest quality class.

Moreover, the enlargement is the only option, because the forest types corresponding to the plantations are coherent with the natural potentiality of the territories and cannot be changed.

TABLE V. FOREST TYPES IN THE TEN CASE STUDIES.

Type Lab[a]	Forest Types	Case Study									
		1	2	3	4	5	6	7	8	9	10
1	Oak-Hornbeam wood of the lowlands			*							
14-15	Oak-Elm wood (also including the Black Alder variant)				*			*			
172	Black Alder wood of gulley			*							
173	Typical Black Alder wood		*		*						
177	Willow wood of bank	*		*	*			*		*	
183	White Poplar formation	*									
188	Pure *Robinia pseudoacacia* wood	*	*	*	*						
189	Mixed *Robinia pseudoacacia* wood		*	*	*	*	*	*			*

a. According to ERSAF database [16]

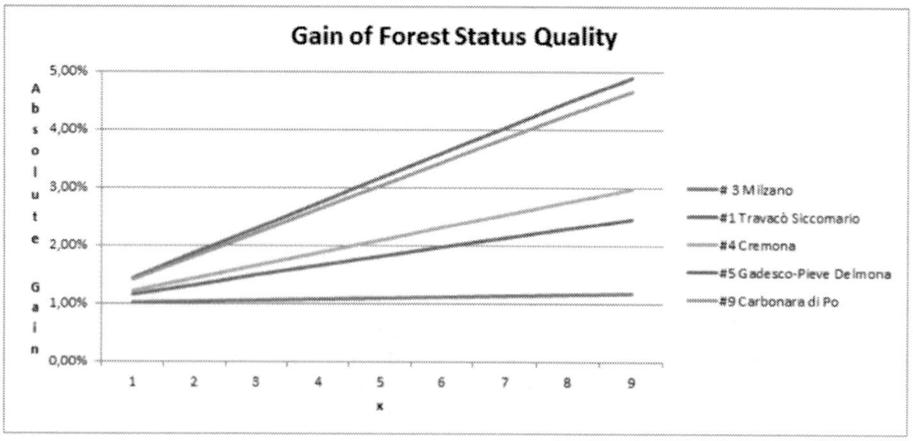

Figure 10. The absolute gain of FSQ' over FSQ, for low-increasing cases.

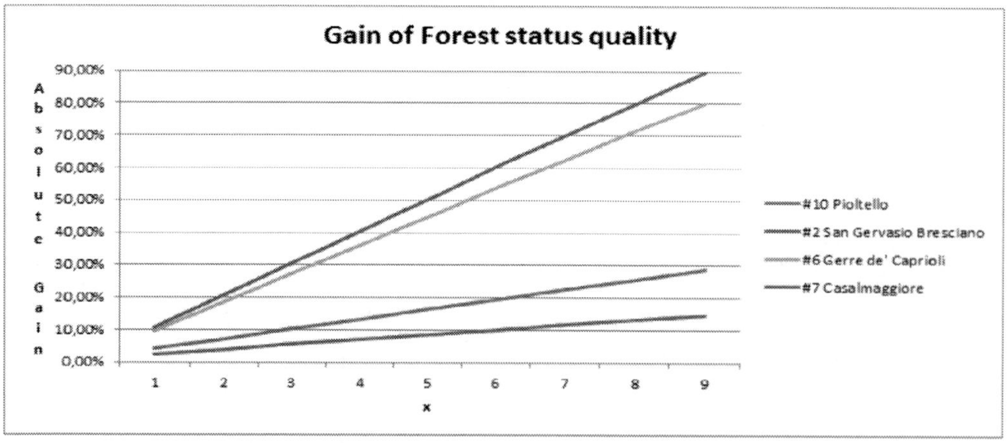

Figure 11. The absolute gain of FSQ' over FSQ, for high-increasing cases.

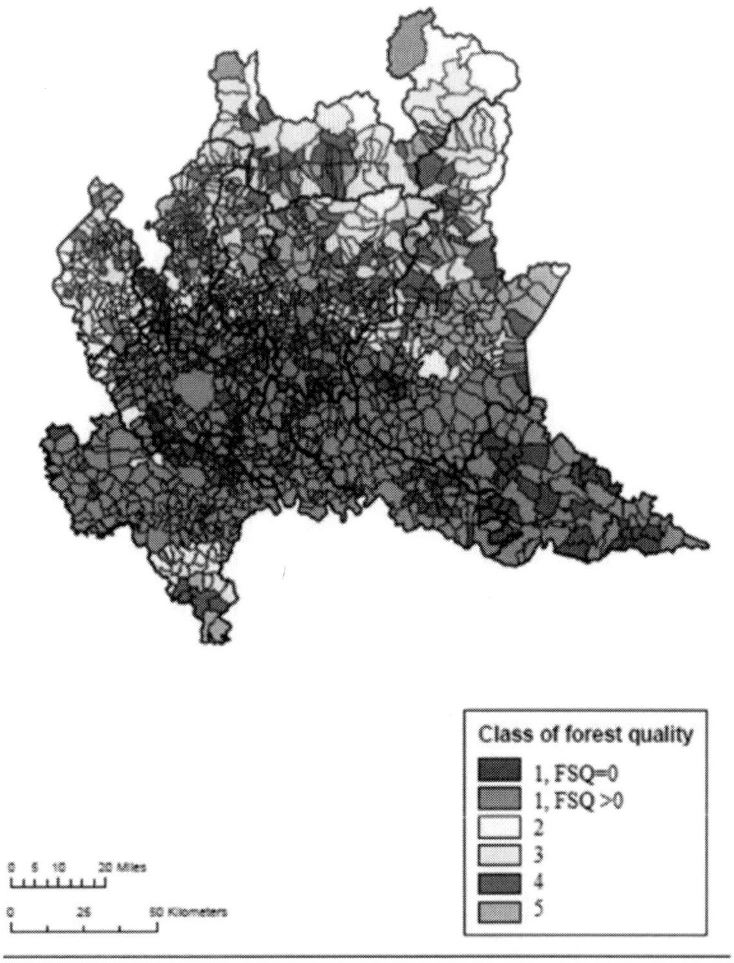

Figure 12. The map of the FSQ indicator on the Region Lombardy.

B. The use of FSQ'indicator for conservation policies

Developing biodiversity conservation planning tools is a big challenge for scientists [23][24]. The two main points a conservation policy should focus on are [25]:

- to identify what are the sites where a *quantitative* target for biodiversity conservation has to be achieved and
- to measure the contribution that these sites make to biodiversity conservation.

We think that the new formulation of the Forest Status Quality Indicator, i.e., the expression of FSQ' (according to (3)) can be a valid help in the achievement of these goals. First of all, the term *quantitative target* is well expressed by an indicator of type B [26], or performance indicator, such as the FSQ' is. In fact, the FSQ' indicator, compares the current conditions of the forest status quality to a specific set of reference conditions (the metric of Table I, which is the same for the original definition of FSQ and for FSQ').

In order to identify the sites where high-impact policy of restoration and/or requalification of forest are preferable, we can use the value of FSQ in order to identify the municipality, i.e., the sites, where the project and realization of plantation can be a valid response to conservation issues. For this purpose, we can consider the municipalities whose FSQ value (according to (1)) is in the class 1 (unsatisfactory, see Table I), i.e., the FSQ value is less than the first range threshold of 0.9. We have computed the values of FSQ for all the 1544 municipalities of Region Lombardy (we recall that this value is computed without the plantation contribution, only natural forest are considered.)

In Figure 12, we have shown the visual map of Region Lombardy; besides the five classes, the first class is split in two colors: in black, to highlight the municipalities of class 1 with FSQ = 0, and in red, the municipalities of class 1 but FSQ > 0. By looking to the map, conservation policy maker should best individuate sites for plantation restoration. The situation in Lombardy is very worrying: a percentage of 54,3% of the municipalities are of class 1, with almost 13% of the municipalities has a FSQ value equal to zero. The distributions of the municipalities for each class of forest quality (according to Table I) is shown in Figure 13.

TABLE VI. THE FOREST TYPES FOR THE TEN CASE STUDIES.

Case study	Forest type corresponding to the plantation	Area of the forest plantation (in ha)
1	Oak-Elm wood	42
2	Mix between Oak-Elm wood, Oak-Hornbeam wood of the lowland, eastern variant, and Oak-El wood, variant with shrubs	37
3	Mix between Oak-Elm wood, Oak-Hornbeam wood, eastern variant, and Oak-El wood, shrubby variant	3
4	Oak-Elm wood	93
5	Oak-Elm wood	4
6	Oak-Elm wood	100
7	Oak-Elm wood	246
8	Oak-Hornbeam wood of the lowland, eastern variant	40
9	Oak-Hornbeam wood of the lowland, eastern variant	15
10	Oak-Hornbeam of the lowland	14

TABLE VII. THE GAIN OF THE PLANTATION, FOR EACH CASE STUDY.

Case study	FSQ (without plantation)	FSQ' (with plantation)	Absolute gain (formula (6))
1	0.056634801	0.228433	4.03%
2	0,022777227	0,336234	14.8%
3	0,138502546	0,163295	1.18%
4	0,060074294	0,151868	2.53%
5	0,011700202	0,02501	2.14%
6	0,014784877	0,925521	62.6%
7	0,003876087	0,271513	70.05%
8	0	0,135032936	NaN
9	0,024306075	0,113267	4.66%
10	0.003438496	0.0775	22.54%

Once we have individuated the sites where conservation policy should be realized, we can measure the effect of the gain of the forest status quality due to new plantations by computing the corresponding value of FSQ', exactly as we have shown for the ten case studies. The absolute gain according to (6) is a good indicator of the efficacy of the conservation policy, as it expresses the improvement of forest quality due to the new plantation realized in the territory under investigation.

Suppose, for example, that a municipality has a current value of FSQ below the first threshold, and we want to reach at least the class number 2. By applying the definition of the metric, this means that FSQ' must be at least greater than 0.9. Supposing to choose a high quality forest for plantation, namely a forest showing the highest value for the naturalistic components (3,3,3), by substituting FSQ' = 0.9 and x = 9 in (6):

$$0.9/FSQ = 1 + 9 * A_p/[S*FSQ] \qquad (8)$$

we can derive the minimum area of the forest plantation A_p to reach the goal:

$$A_p = S * (0.9-FSQ)/9 \qquad (9)$$

In this way, we can use the naturalistic components, the FSQ and FSQ' indicators to help policy makers to determine *where* and *how* perform a possible policy of forest requalification by new plantation settlements. However, the choice of the forest type in the area of plantation cannot primarily follow the criterion of the highest value for the triplet of the naturalistic components, but the one of the *coherence* of the forest type with the natural potentiality and ecological conditions of such area. Subsequently, the area of the plantation sufficient to reach our goal will be calculated, as in (9). To better understand this concept of *coherence*, we can consider the example of the plain rivers, where forests will not reach the highest values for the naturalistic component, due to the human and natural disturbance. For this reason, policy makers have to be aware of the necessity of choosing wide areas for the plantations.

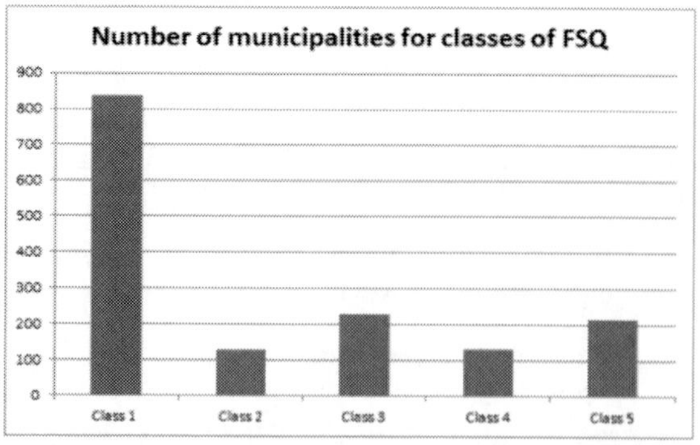

Figure 13. The distribution of all the municipalities of Region Lombardy, in each class of forest quality, according to the metric (See Table I).

IV. Conclusion and Future Work

In this paper, an application of the innovative indicator for forest quality assessment has been proposed with the aim to support planning policies for forest conservation. Despite the fact that it is an experimental research model, the results suggest that it could be used in real world purposes and cases. Actually, the analysed case studies confirm that our methodological approach is realistic in evaluating the forest quality. Moreover, it can give useful information about (a) the choice of forest types to enhance and (b) the surfaces necessary to plantations, if at least a satisfactory level of forest quality is pursued. The computation of such indicator on the territory of Region Lombardy is particularly challenging due to the following features: Region Lombardy ranks first in Italy for the population and for the number of local municipalities, second for population density, and fourth for area. It is also the most invaded Region by non-native species, which represent the 16.9% of the total vascular flora [27]. Furthermore, in Lombardy more than 800 municipalities show an unsatisfactory level of forest quality, and they are more concentrated in the Po Plain. Such Plain is a very polluted area, with values of PM2.5 (mg/cubic meter) between 26 and 35, and its greatest city, Milan, showing values between 36 and 69 [28]. Here, we firmly believe that serious policies of forest restoration are crucial, not only for biodiversity conservation, but also for the human health, due to the important role of forests in pollution control [29].

The present application of the FSQ, together with the previous contributions on the same topic [1][2][3], highlights the strength of the synergy between the botanical and computer science competences in addressing problems about environmental topics and in providing indicator based solutions for policy makers and planners.

References

[1] S. Assini and M. G. Albanesi, "A New Improvement of the Naturalistic Indicator of Forest Quality. An Italian Case Study," Proceedings of the Seventh International Conference on Bioenvironment, Biodiversity and Renewable Energies (BIONATURE 2016), Lisbon (Portugal), 26-20 June 2016, pp. 7-10, Copyright (©) IARIA, 2016, ISBN 978-1-61208-489-3.

[2] S. Assini and M. G. Albanesi, "A Naturalistic Indicator of the Forest Quality and its Relationship with the Land Use Anthropentropy Factor," Proceedings of the Sixth International Conference on Bioenvironment, Biodiversity and Renewable Energies (BIONATURE 2015), Rome (Italy), 24-29 May 2015, Copyright (©) IARIA, 2015, pp. 27-33, ISBN 978-1-61208-410-7.

[3] S. Assini and M.G. Albanesi, "A New Biodiversity Composite Indicator Based on Anthropentropy and Forest Quality Assessment. Framework, Theory, and Case Studies of Italian Territory," IARIA International Journal On Advances in Life Sciences, 7 (3-4), 2015, pp. 177-194.

[4] T. Gao, A.B. Nielsen, and M. Hedblom, "Reviewing the strength of evidence of biodiversity indicators for forest ecosystems in Europe," Ecological Indicators, 57, 2015, pp. 420-434.

[5] S. Schindler, K. Poirazidis, and T. Wrbka, "Towards a core set of landscape metrics for biodiversity assessments: A case study from Dadia National Park, Greece," Ecological Indicators, 8, 2008, pp. 502-514.

[6] P. Digiovinazzo, G. F. Ficetola, L. Bottoni, C. Andreis, and E. Padoa-Schioppa, "Ecological thresholds in herb communities for the management of suburban fragmented forests," Forest Ecology and Management, vol. 259, 2010, pp. 343-349.

[7] D. A. Saunders, R. J. Hobbs, and C. R. Margules, "Biological consequences of ecosystem fragmentation: a review," Conservation Biology, vol. 5, 1991, pp. 18-32.

[8] O. Honnay, K. Verheyen, and M. Hermy, "Permeability of ancient forest edges for weedy plant species invasion," Forest Ecology and Management, vol.161, 2002, pp.109-122.

[9] W. F. Laurance et al., "Ecosystem decay of Amazon forest fragments: a 22-year investigation," Conservation Biology, vol. 16, 2002, pp. 605-618.

[10] Z. Dzwonko and S. Loster, "Species richness of small woodlands on the western Carpathians foothills," Vegetatio, vol. 76, 1988, pp. 15-27.

[11] Z. Dzwonko and S. Loster, "Species richness and seed dispersal to secondary woods in southern Poland," Journal of Biogeography, vol. 19, 1992, pp. 195-204.

[12] D. R. Patton, "A diversity index for quantifying habitat edge," Wildlife Society Bulletin, vol. 3, 1975, pp. 171-173.

[13] O. Honnay, M. Hermy, and P. Coppin, "Effects of area, age and diversity of forest patches in Belgium on plant species richness, and implication for conservation and restoration," Biological Conservation, vol. 87, 1999, pp. 73-84.

[14] F. Sartori and S. Assini, "First evaluation of the regional afforestation actions of the lombardy plain," Transl. "Prime valutazioni sulle opere regionali di riforestazione della pinaura lombarda," «NATURA BRESCIANA» Ann. Mus. Civ. Sc. Nat., Brescia, 36, 2009, pp. 191-196.

[15] S. Assini and F. Sartori, "Vegetation and restoration – Monitoring of the actions: the great plain forests of the Lombardy Region. Analyses and perspectives," Transl. "Vegetazione e riqualificazione – Monitoraggio degli interventi: le grandi foreste di pianura della Regione Lombardia. Analisi e prospettive," Quaderni di Tutela del Territorio, Regione Piemonte, 3, pp. 57-65, 2009.

[16] Regional Regional Agency for Services to Agriculture and Forests, Transl. Ente Regionale Servizi Agricoltura e Foreste, "Map of Forest Types of Lombardia," Transl. "Carta dei tipi reali forestali della Lombardia," http://www.geoportale.regione.lombardia.it/galleria-mappe, page 3 [retrieved: May, 2017].

[17] C. Andreis and F. Sartori, eds, "Forest vegetation of Lombardia. Phytosociological classification," Transl. "Vegetazione forestale della Lombardia. Inquadramento fitosociologico," Archivio Geobotanico, vol. 12-13, 2011, pp. 1-215.

[18] QuantumGIS software, http://www.qgis.org/en/site/ [retrieved: May, 2017].

[19] R. C. Gonzales and R. E. Woods, Digital image processing, Pearson Prentice Hall, 2008, Chapter 9, Morphological image processing.

[20] ISTAT, On line database of Territorial and Cartographic information, http://www.istat.it/it/strumenti/territorio-e-cartografia [retrieved: May, 2017]

[21] Matlab software, https://it.mathworks.com/help/images/ [retrieved: May, 2017].

[22] R. del Favero, ed., "The forest types of Lombardy," Transl. "I tipi forestale della Lombardia", CIERRE Edizioni, pp. 506, 2002.

[23] D. Simberloff, "The role of science in the preservation of forest biodiversity," Forest Ecology and Management , vol. 115, 1999, pp. 101-111.

[24] L. Failinga and R. Gregory, "Ten common mistakes in designing biodiversity indicators for forest policy," Journal of Environmental Management, vol. 68, 2003, pp. 121–132.

[25] Sahotra Sarkar et al., "Biodiversity Conservation Planning Tools: Present Status and Challenges for the Future," Annu. Rev. Environ. Resour, vol. 31, 2006, pp. 123–59. doi: 10.1146/annurev.energy.31.042606.085844.

[26] E. Smeets and R. Weterings, "Environmental indicators: typology and overview," European Environment Agency Technical Report n. 25/1999 http://www.eea.europa.eu/publications/TEC25 [retrieved: May, 2017]

[27] L. Celesti-Grapow et al., "Non-native flora of Italy. Species distribution and threats," Plat Biosystems, vol. 144(1), 2010, pp. 12-28.

[28] Global ambient air pollution, World Health Organization, http://maps.who.int/airpollution/ [retrieved: May, 2017]

[29] Forest Ecosystem Services, Europen Commission JRC, http://forest.jrc.ec.europa.eu/activities/forest-ecosystem-services/ [retrieved: May, 2017].

Inertial Sensors-based Lower-Limb Rehabilitation Assessment: A Comprehensive Evaluation of Gait, Kinematic and Statistical Metrics

Salvatore Tedesco[1], Andrea Urru[1], James Peckitt[2], Brendan O'Flynn[1]
[1]WSN Group, Micro & Nano Systems - Tyndall National Institute
[2] Mardyke Arena Clinic
University College Cork
Cork, Ireland
e-mail: salvatore.tedesco@tyndall.ie

Abstract— **Analysis of biomechanics is frequently used in both clinical and sporting practice in order to assess human motion and their performance of defined tasks. Whilst camera-based motion capture systems have long been regarded as the 'Gold-standard' for quantitative movement-based analysis, their application is not without limitations as regards potential sources of variability in measurements, high cost, and practicality of use for larger patient/subject groups. Another more practical approach, which presents itself as a viable solution to biomechanical motion capture and monitoring in sporting and patient groups, is through the use of small-size low-cost wearable Micro-ElectroMechanical Systems (MEMs)-based inertial sensors. The clinical aim of the present work is to evaluate rehabilitation progress following knee injuries, identifying a number of metrics measured via a wireless inertial sensing system. Several metrics in the time-domain have been considered to be reliable for measuring and quantifying patient progress across multiple exercises in different activities. This system was developed at the Tyndall National Institute and is able to provide a complete and accurate biomechanics assessment without the constraints of a motion capture laboratory. The results show that inertial sensors can be used for a quantitative assessment of knee joint mobility, providing valuable information to clinical experts as regards the trend of patient progress over the course of rehabilitation.**

Keywords- Inertial Sensors; Wearable Microsystems; Signal Processing; Data Analytics; Lower-Limb Rehabilitation; Motor Performance.

I. Introduction

This paper is an extended presentation of [1], a study on the qualitative assessment of progress during rehabilitation via wearable inertial sensors, first published at Global Health 2016.

Biomechanics is the science related to the study of the internal and external forces acting on the human body and the effects produced by them [2]. In particular, one aspect of biomechanics analysis is the study of human locomotion and of the forces causing movement and human kinetics.

This analysis is frequently used in both clinical and sporting practice by clinicians and can play a crucial role in athletes' performance enhancement, injury prevention, and effective rehabilitation. More specifically, in the latter case, it is essential to track patient progress, and consequently to tailor patients-oriented rehabilitation programs, through the accurate assessment of human motion during the performance of clinically defined tasks.

A common example of the technology regarded as being the 'Gold-standard' of quantitative movement analysis is shown by camera-based motion analysis systems (such as the ones provided by Vicon [3], Optitrack [4], or Codamotion [5]) which, during formal gait analysis by rehabilitation professionals, can help to ascertain measurements of Temporal (Time) and Spatial (Distance) characteristics associated with gait parameters. This enables clinicians to identify gait deviations in paediatric and amputee populations, in screening elderly people at risk of falling, to objectively monitor a patient's progress, and to help determine the efficacy of surgical and therapy interventions [6-9].

While camera-based motion capture systems achieve very high performance, their application is not without limitations as regards potential sources of variability in measurements, relatively high costs of instrumentation including access to specialist motion labs, as well as practicality of application for larger patient/subject groups, as discussed by Chau et al. in [10]. Similar drawbacks have been demonstrated for other accurate, clinical grade, gold-standard measurement systems, such as marker less video [11], instrumented treadmills [12], walkway contact mats [13], or force platforms [14].

From a clinical perspective, observational forms of clinical gait analysis frequently forms the corner stone of patient knee joint assessment, and is typically used in parallel with manual clinical assessment techniques. These include stress-testing evaluation of joint laxity, range of movement (ROM), and manual and/or isokinetic strength assessment, as well as contextual subjective patient questionnaires, such as Knee Injury and Osteoarthritis Outcome Score (KOOS), Oxford Knee Score (OKS), Tegner Lysholm Knee Scoring Scale, International Knee Documentation Committee (IKDC), and Western Ontario & McMaster Universities Osteoarthritis Index (WOMAC) [15][16], which employ a high degree of patient involvement.

However, the use of observational gait analysis (non-empirical assessments), even when utilised by experienced clinicians, may not be adequate or sensitive enough to detect subtle clinical pathological changes in movement following knee surgery [8] [9].

An alternative approach, which has been explored as a more practical and viable solution to biomechanical motion capture and monitoring in sporting and patient groups, involves the use of small-size low-cost wearable inertial sensors [17].

There has been a wide variety of work presented in literature on inertial sensors applied to biomechanics. These have been typically adopted for monitoring the lower-limbs during rehabilitation or tele-rehabilitation and used to objectively assess the performance of impaired subjects throughout the process, in particular following knee injuries. However, most of those investigations were focused on a one-time assessment rather than a longitudinal evaluation over several weeks looking at change in gait over longer periods.

The aim of the present work is to evaluate motor performance during lower-limb rehabilitation targeting activities normally assigned by clinicians for at-home rehabilitation. The work also addresses the assessment of the performance of body-worn kinematic sensors in a rehabilitative context, given their well-known potential in accurately extracting parameters that inform qualitatively and quantitatively movement parameters.

Informed by clinical partners involved in the development and validation of the system, the study investigates and derives a number of features and metrics, related to gait and kinematic characteristics and statistical analysis. These data sets will be analysed to establish which of them are the most sensitive and helpful to determine changes in motor capacity over a longitudinal study of nine months. The data sets acquired will help develop, in future works, better models for objectively estimating the conditions and the motor performance of adults involved in lower-limb at-home rehabilitation following knee injuries.

The derived outcome of this model will be analyzed by clinicians and sport scientists to gain a comprehensive picture of patients' condition and provide more targeted medical feedback.

The analysis is carried out by using a wearable inertial sensing system developed at the Tyndall National Institute, consisting of two sensors per limb, which is able to provide a complete biomechanics assessment for a series of scripted activities, based on best clinical practice.

The present work is organized as follows. Relevant recent works are discussed in Section II while the description of the hardware and of the test protocol used during the data collection are described in Section III. The methodology behind the feature selection is illustrated in Section IV. The obtained results are shown in Section V and exhaustively analyzed and discussed. Finally, conclusions are drawn in the last section.

II. Related Works

Inertial sensors are generally used in devices to measure velocity, orientation, and gravitational force, and are used in a great number of applications. These include industry quality control, robotics, navigation systems, sports, augmented reality systems, and so on [18-21]. Biomechanics, in particular, has achieved significant progress from the adoption of this technology [22].

More specifically, with regards to gait analysis, accelerometers and gyroscopes have been used worn on the lower-limbs to obtain gait parameters [23-26], which can be derived by the integration of linear acceleration or angular velocity, after the correct identification of the beginning and the end of each gait cycle. Interesting innovations have been proposed in [27] by L. Atallah et al. in the development of an ear-worn sensor for gait monitoring, or by S. Kobashi et al. [28] who included magnetic sensors in combination with inertial sensing to estimate knee joint angle in three dimensions.

This technology has not only been taken into account for healthy subjects. Such systems have also been used for the detection of pathologic conditions, discriminating clinical indications between symptomatic and asymptomatic subjects in a number of diseases. Conditions which have been investigated in this way include cerebral palsy [29], hemiplegia [30-31], Parkinson's [32], dementia [33], old age [34], and Anterior Cruciate Ligament (ACL) injury [35].

In recent years, researchers have increasingly been investigating the use of inertial sensors in gait rehabilitation. Most of the studies in this area focused on the implementation of a lower-limb monitoring system for remote rehabilitation or tele-rehabilitation [36-38], or in the performance assessment of specific rehabilitation exercises [39-46]. In the latter case, machine learning techniques are adopted to discriminate between the correct and incorrect execution of recommended exercises. The most advanced techniques can also highlight if multiple incorrect postures are present while performing the test. Typically, a number of body-worn sensors are used for the classification. However, there is a significant body of research in the investigation of the efficacy of the adoption of only one sensor [47-50]. This can be obtained also through the adoption of specific biomechanical models related to particular exercises.

Immersive virtual reality, computer games, or visualization tools are recently being developed in order to enhance patients' adherence to the rehabilitation program and enhance motivation [51-53]. The accuracy and reliability of those inertial sensors and biofeedback-based rehabilitation systems have been shown in [54-55].

To date, however, few studies considered the quantitative assessment of patients' lower-limb performance via body-worn sensors during the complete rehabilitation process. This task is particularly challenging as it consists of isolating the gradual changes in movements due to recovery and improvement despite the presence of a multitude of sources of variability. The temporal and spatial sources of intra- and inter-variability (e.g., dissimilarities in repetitions of an exercise when performed by one subject, and dissimilarities between different subjects, respectively) are already evident in healthy subjects, given the different characteristics of gender, age, height, and weight, etc…, and are even more significant in patients following rehabilitation, due to different levels of pain, fatigue, and possible compensations.

For instance, Lin et al. [56] estimated the joint angles with a novel extended Kalman filter on 14 exercises performed by a cohort of elderly patients monitored from the first day of admission until discharge, with the average patient's treatment lasting 5.7 days.

Similarly, Field et al. [57] investigated the gradual changes of motion with new proposed metrics, by monitoring the symmetry between the left and right sides of the body for 14 subjects over repeated rehabilitation exercises in a period of 12 weeks. However, despite the completeness of the method, the study required the patient to wear a motion capture suit consisting of 17 sensors, which is cumbersome and not feasible for at-home rehabilitation.

Finally, in [58], a novel machine learning technique has been proposed that estimates the continuous measurement of patient improvement which is capable of handling a variety of rehabilitative exercises. The approach was tested by adopting two wearables sensors on thigh and shank on clinical data collected on 18 elderly patients involved in rehabilitation following hip and knee replacements for a range of 4-12 days.

However, the main limitations of those studies are related to the short period for data collection which explores only the initial part of the rehabilitative process without considering the long-term effects or the pre-surgery conditions. Another limitation is the need of a large and specific initial dataset on which the machine learning method has to be trained. Finally, the lack of definition of the variation of the selected features throughout the analysis period and their impact on the final outcome is a constraint in using this approach.

The present study, as an extended version of [1], will analyze the data collected from a patient in pre/post-surgery conditions for an overall period over nine months, with the twofold aim of:

1. investigating, through body-worn inertial sensing, the effects of rehabilitation over different periods, also in the long-term, monitoring patient's progress,
2. understanding which parameters, taken from a wide range of features described in literature, and informed by clinical inputs, can be the most beneficial and sensitive for clinicians when monitoring patients in the course of lower-limb rehabilitation.

III. Hardware and Protocol for Data Collection

The biomechanical monitoring system consists of two Tyndall Wireless Inertial Measurement Units (WIMUs) [59] per leg, each one with 3D accelerometer and gyroscope (@ 250 Hz) and Bluetooth Low-Energy for wireless communication (Figure 1). WIMUs have been attached to the anterior tibia, 10 cm below the tibial tuberosity, and to the lateral thigh, 15 cm above the tibial tuberosity using surgical adhesive tape.

Figure 1. Tyndall Wireless Inertial Measurement Unit (WIMU)

The rehabilitation exercises (or scenarios) considered are walking, half squat, hamstring curl, and flexion/extension – defined by physiotherapists as good indicators of rehabilitation progress. These are described as follows:

- In the walking scenario, the subject walks on a treadmill, which is operated at defined speeds (3-4-6 km/h) for approximately one minute per test.
- In the half squat scenario, the subject stands with the feet shoulder's distance apart and arms crossed on the chest. Keeping the chest lifted, the hips are lowered about 10 inches, planting the weight in the heels. The body is then brought back up to standing by pushing through the heels.
- In the hamstring curl scenario, the subject stands and bends the knee raising the heel toward the ceiling as far as possible without pain, relaxing the leg after each repetition. This is repeated on both legs.
- In the flexion/extension scenario, the subject lies supine on the floor and bends the knee raising it toward the chest as far as possible without pain, relaxing the leg after each repetition. This is repeated on both legs.

The system has been tested with an impaired subject. The impaired subject is a female athlete, age: 44, height: 161 cm, and weight: 52 kg, with good general health status, with a history of knee injuries and surgery (reconstructed anterior cruciate ligament in the left leg following a sporting injury). The tests were carried out during the course of the rehabilitation program, e.g., starting 1 month before surgery and finishing 7 months after surgery. Overall, the subject has been evaluated through three periods: once in pre-surgery conditions (e.g., 1 month before surgery), then 6 times in a range of 20 weeks starting one month after surgery (namely short-term post-surgery), and finally once 3 months after the last data capture (e.g., during long-term post-surgery period).

A number of repetitions has been collected for each scenario, so as to provide an accurate picture of the overall conditions, and each scenario was evaluated during almost every data capture. The hamstring curl scenario as well as the walking test at 3 and 4 km/h was performed at every session.

Similarly, the flexion/extension test was always recorded except in the pre-surgery session due to subject's impairment of movement. For the same reason, half squat and walking at 6 km/h were not recorded in the first 2 sessions of the short-term post-surgery period.

IV. FEATURES SELECTION

The metrics taken into account for the patient's assessment are divided into three main categories: statistical metrics, gait characteristics, and time-domain kinematic measurements.

i. Statistical Features

This category takes into account various well-known statistical features extrapolated from the time-domain. Those variables are applied on every segmented walking stride/exercise repetition for both legs performed during the sessions. The selected features are described below:

- Mean, standard deviation, variance, skew, kurtosis, root mean square, signal magnitude area, energy (given by the integration over the repetition of the squared absolute signal) of the acceleration and angular velocity magnitudes;
- Mean, minimum, maximum, median, standard deviation, coefficient of variation (CV), peak-to-peak (PP) amplitude, and root mean square of the x-, y-, and z-axis of the acceleration and angular rate signals;
- Autocorrelation on the x-, y-, and z-axis of the acceleration and angular rate signals measured taking into account all the repetitions/strides in a session as a whole and not separately.

All those features are calculated for each of the 4 sensors used for data collection.

ii. Gait Characteristics

Well-known gait measures [1] are calculated from the data recorded by the inertial sensors attached on the shanks as follows:

- Gait cycle time (GCT), which is the time-interval between two consecutive toe-offs of the same leg;
- Stance phase, defined as the weight-bearing phase of the GCT in which the body is supported, and is expressed by the difference between a heel-strike and the following toe-off of the same leg;
- Relative stance phase, the ratio between stance phase and GCT;
- Swing phase, which is the non-weight bearing phase of GCT, and is expressed by the difference between a heel-strike and the previous toe-off of the same leg;
- Relative swing phase, the ratio between the swing phase and the GCT;
- Double support, the time-interval when both feet are in ground contact;
- Stride length, indicated as the distance between two successive placements of the same foot, computed as the total trajectory on the sagittal plane made by the sensor attached to the shank. The approach is based on a double integration of inertial data collected in a stride. The integration process is reset at the end of each stride;
- Stride speed, computed by the ratio between stride length and GCT;
- Clearance, defined as the maximum height reached by the sensor during the swing phase and obtained by the vertical displacement calculated to establish the stride length during the swing phase.

For each variable, also the related CV has been extrapolated, and consequently, also the associated symmetry, defined for each specific parameter as

$$Symmetry_{parameter} = \frac{|CV_{parameter\_left} - CV_{parameter\_right}|}{\mu(stride\_speed)} \quad (1)$$

where $\mu(\text{stride\_speed})$ is the average of the measured stride speed for that specific session.

Moreover, the Balance Index (B.I.), as shown in [31], expressed as the absolute ratio between the difference of the left and right leg's values of a specific gait parameter and their sum, was obtained for all the variables.

This information is obtained for both legs for the walking scenario only.

iii. Kinematic Variables

Finally, the third category is related to kinematic metrics, occasionally adopted for gait analysis, but that can provide useful information on the movement analysis, and have been proposed in recent works in literature. Those metrics are as follows:

- Knee Range of Motion (ROM), defined as the peak-to-peak amplitude of the knee joint angle during a repetition;
- Regularity [60], e.g., the ratio between the unbiased autocorrelation coefficient at the first dominant period and the coefficient at the second dominant period, both measured taking into account all the repetitions within the same analyzed scenario;
- Range of Angular Velocity (RANG) [61-62]: the difference between the minimum and the maximal value of the angular velocity magnitude within each repetition;
- Jerk-based smoothness measures, where the jerk is the rate of change of the acceleration in a repetition. Several jerk-based metrics have been proposed, including integrated squared jerk (ISJ), mean squared jerk (MSJ),

cumulative square jerk (CSJ), root mean square jerk (RMSJ), mean square jerk normalized by peak speed (N_MSJ), integrated absolute jerk (IAJ), mean absolute jerk normalized by peak speed (N_MAJ), and dimensionless square jerk, whose mathematical definitions are shown in [63];

- Vertical acceleration [64], defined as the maximum value over a repetition of the difference between the acceleration magnitude (filtered with a 2nd order Butterworth low-pass filter with cut-off frequency 3 Hz) and the gravitational force;
- Vertical velocity [64], defined as the integration over a repetition of the difference between the acceleration magnitude (filtered with a 2nd order Butterworth low-pass filter with cut-off frequency 3 Hz) and the gravitational force;
- Fluency [64], e.g., the integration over a repetition of the absolute difference between the raw and the filtered x, y, and z-axis acceleration signals. Again the filter used is a 2nd order Butterworth low-pass filter with cut-off frequency 3 Hz;
- Stability [65], defined as the dynamic time warping of the x-, y-, and z-axis of the acceleration and angular rate signals measured at two consecutive repetitions/strides, then averaged based on all the repetitions present in a test session;
- Kinetic Value (KV) [66], defined as the squared integral of the magnitude of the acceleration signal over a repetition, multiplied by *m/2*, where *m* is the subject's body weight.

All those features in this class are calculated for each of the 4 sensors used for data collection.

The data analysis is implemented off-line over the data collected using a commercial software package (MATLAB R2015a, The MathWorks Inc., Natick, MA, 2015) [67]. Each repetition/stride was visually segmented.

V. Results and Discussion

In each session, each scenario was divided in two separate tests (both logged for 60 sec), and in each of the two tests a series of repetitions have been carried out by the subject. The overall number of repetitions recorded for all the sessions was: 184 hamstring curls (92 left and 92 right), 134 flexion/extensions (67 left and 67 right), 66 half squats, 478 strides for both legs when walking at 3 km/h, and similarly 544 strides when walking at 4 km/h, and 512 strides when walking at 6 km/h. For each test, the features described in Section IV were extrapolated and compared among the different sessions, in order to provide a clear understanding of which metrics can be more valuable during rehabilitation. The results are summarized in different tables (collected in the Appendix, which is available as a PDF file at the following link: https://www.tyndall.ie/contentfiles/Tables_IARIA.pdf) where each table consists of 9 columns given by all the sessions in which the test were carried out. Each session is divided in three sub-columns: one for indicating the results for the left leg, one for the right leg, and the last one for the mean difference (expressed in percentage) between left/right values (considering the right leg values as references), the Pearson's r coefficient and the p-value between the left/right leg results, calculated when appropriate. The sub-columns related to left and right results are merged in case of gait metrics which in their calculation takes into account aspects from both legs. Due to the fact that the subject could not perform all the scenarios during each session, not all the columns in the tables are filled. Moreover, owing to malfunctioning issues during data recording, results from the right leg in the hamstring curl scenario on the first session are not available.

Given the number of tables extrapolated from the data, only a small subset of those tables are included in this paper for clarification, while full details are shown in the Appendix.

The mean difference, in particular, is an important estimator of the dissimilarities between the two legs which, in an ideal case, should be close to zero in any case for a healthy unimpaired subject. Given its definition, negative values of the mean difference indicate that results for the right leg are larger compared to the left one, and *vice versa* for positive values.

Finally, in order to have the same reference system for both WIMUs worn on the same leg, the method proposed by Seel et al. [68] has been adopted to virtually rotate around an axis the raw inertial data recorded on the shank. As a result, for all the WIMUs involved, the x-axis represents the medio-lateral axis, the y-axis is the anteroposterior one, while the z-axis is the vertical axis. Thus, the plane y-z represents the sagittal plane.

Results from data analysis, divided in three categories, are described below.

i. Gait Metrics

The analysis performed on the gait characteristics is summarized in Tables I-II-III available in the Appendix and in this paper, which represents the walking test at 3, 4, and 6 km/h, respectively. All the gait features are indicated in those tables, with B.I. and symmetry calculated for each metrics and shown together in the same row.

Considering the gait results at 3 km/h, there are several metrics which may be beneficial for showing the patient's progress during rehabilitation. As an example, even though the p-value related to the GCT in all sessions is above 0.05 (e.g., there is no statistical difference between the two legs), the calculated mean difference between left/right for this parameter has a clear convergence to zero during the monitoring phase, and the same trends are evident also for the mean difference of the CV, B.I., and symmetry associated to GCT. Similar considerations can be observed in the results provided by the CV obtained for the stance phase, swing phase, and double support. In particular, the mean difference associated to the CV of the stance phase shows a convergence after the second session, as it is evident a clear increase of the mean difference between the first two sessions (the pre-surgery session, and the first test after surgery), due to the early stage of the recovery process post-surgery.

Again, even though the p-values for the stride length is constantly below 0.05 in the different sessions (e.g., there is always a significant difference between the two legs), the Pearson's coefficient continuously increases, while the mean difference tends to zero. The mean difference of the CV and the B.I. of the stride length shows a comparable convergence as well, also for what concerns the dissimilarities between the pre-surgery session and the first post-surgery one. Similar considerations may be drawn for the stride speed, also explained by the uniformity of the results given by the GCT.

The gait scenario at 4 km/h shows similar trends for the GCT-related parameters, and the mean difference of the CV related to the stance phase and double support, with an evident convergence towards zero and a strong disparity in the first two sessions. However, no particular correlation is clear for the stride length values. On the other hand, clearance may highlight some significant information. P-values are always lower than the statistical threshold (0.05), except that in the last session recorded seven months after surgery, and its mean difference and the B.I. measure show a clear convergence toward zero during the rehabilitation assessment, indicating, thus, a certain gained equivalence between the two legs.

Finally, gait results at 6 km/h show significant trends only for what concerns the mean difference of the CV of the double support and clearance-related parameters.

In summary, it is evident how, at slow speed, temporal parameters (especially indicated through their CV) and the stride length values should be considered for assessing differences between left and right legs during rehabilitation, whereas, increasing the speed, clearance seems to gain higher priority than time-related parameters. Only the mean difference of the CV of the double support shows a similar trend at all the tested speeds. Some of the discussed results are illustrated in Figures 2-4. The markers in the figures indicate the mean value obtained for a specific session, while the green line is a reference for zero.

ii. Statistical Features

The statistical features described in Section IV has been calculated for each repetition/stride for all the four WIMUs adopted and for both acceleration and angular rate signals. Results are summarized for all the scenarios in the tables in the Appendix and discussed below.

In the hamstring curl scenario, for example, the thigh is not significantly involved in the movement. Indeed, there are no specific variables showing improvements. Conversely, the shank is more informative. Standard deviation, variance, calculated on the acceleration magnitude, and standard deviation, CV, and peak-to-peak amplitude obtained from the z-axis acceleration all show clear convergences. Moreover, those trends are even more evident from metrics obtained by the angular rate collected on the shank. For instance, mean, standard deviation, variance, level of skew, signal magnitude area calculated from the magnitude signal, and minimum, maximum, standard deviation, and peak-to-peak amplitude on the three axis are all significant variables (Figure 5).

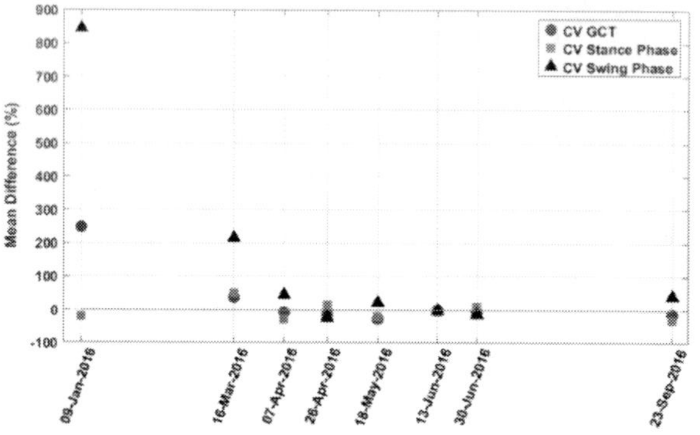

Figure 2. Mean difference for different gait parameters at 3 km/h showing the trends of progress during rehabilitation. CV GCT/stance phase/swing phase are shown

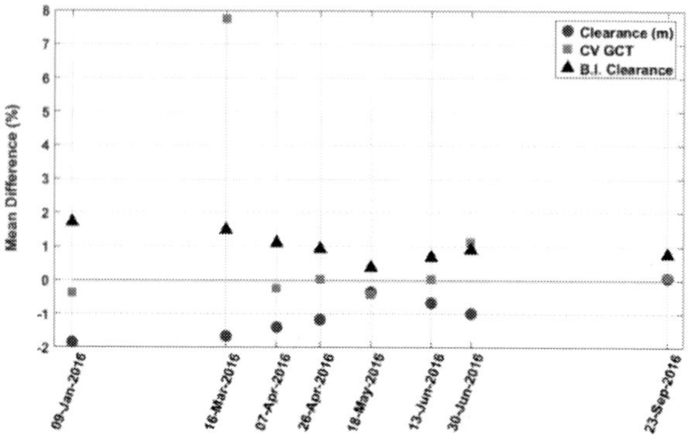

Figure 3. Mean difference and balance index for different gait parameters at 4 km/h showing the trends of progress during rehabilitation. Clearance CV GCT, and B.I. clearance are shown. Each variable is normalized according to its mean value for visualization purposes

In the flexion/extension scenario, the movement requires a higher involvement of the thigh compared to the curl which, thus, present several metrics useful for progress monitoring (Figure 6). Examples are the mean, standard deviation, variance, skewness, kurtosis, signal magnitude area, and energy obtained from the angular rate magnitude, minumum, standard deviation CV, and peak-to-peak amplitude from the acceleration vertical axis, and minimum, maximum, standard deviation, and peak-to-peak amplitude from the angular rate collected around the sagittal axis. There are also numerous metrics related to the shank, such as the mean, standard deviation, variance, skewness, signal magnitude area, and energy obtained from the acceleration and angular rate magnitude, and a number of variables obtained from the single axis of the inertial sensors and associated to the sagittal plane. This higher number of metrics is due to the intrinsic movement as defined by the exercise which occurs almost completely on the sagittal plane and, thus, the information is highlighted around the rotation axis and not divided along the three axes. Indeed, this characteristic is not present in the remaining scenarios.

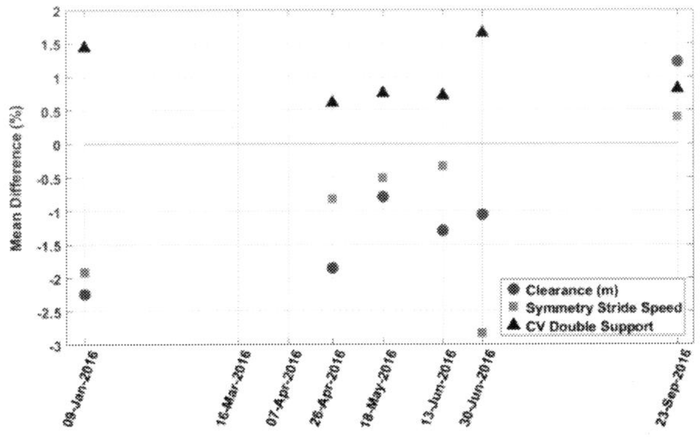

Figure 4. Mean difference and symmetry for different gait parameters at 6 km/h showing the trends of progress during rehabilitation. Clearance, symmetry stride speed, and CV double support are shown. Each variable is normalized according to its mean value for visualization purposes

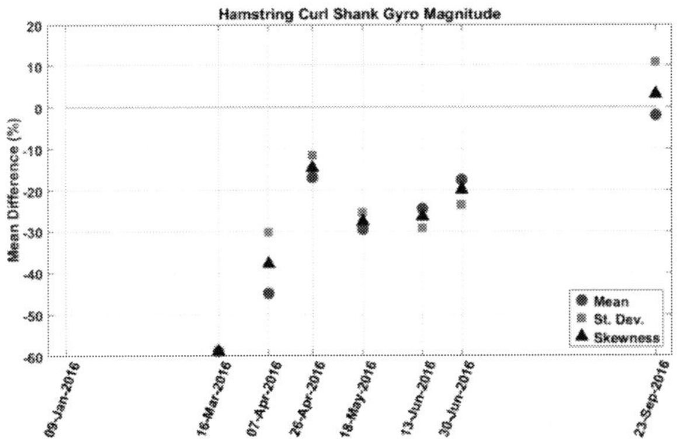

Figure 5. Mean difference for different statistical parameters in the curl scenario showing the trends of progress during rehabilitation. Mean/St. dev/skewness for the gyro shank magnitude are shown

In the half squat scenario only a few metrics are evident on separate axes. For example, the maximum and peak-to-peak value of the acceleration on the y-, z-axis measured on the thigh and shank, and the standard deviation, CV, and peak-to-peak amplitude measured along the sagittal axis on the angular rate of the lower-limbs illustrate a reasonable progress during the rehabilitation period, while the metrics (mean, level of skew, area, energy) calculated from the magnitude of the inertial data on the shank are more informative (Figure 7).

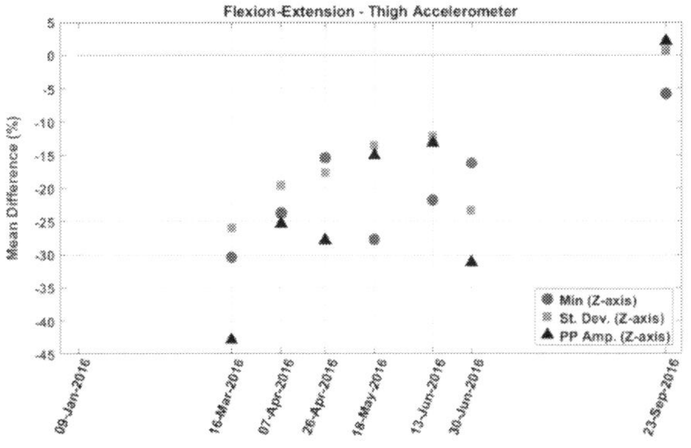

Figure 6. Mean difference for different statistical parameters in the flexion/extension scenario showing the trends of progress during rehabilitation. Minimum/st dev/PP amplitude for the acceleration thigh z-axis are shown

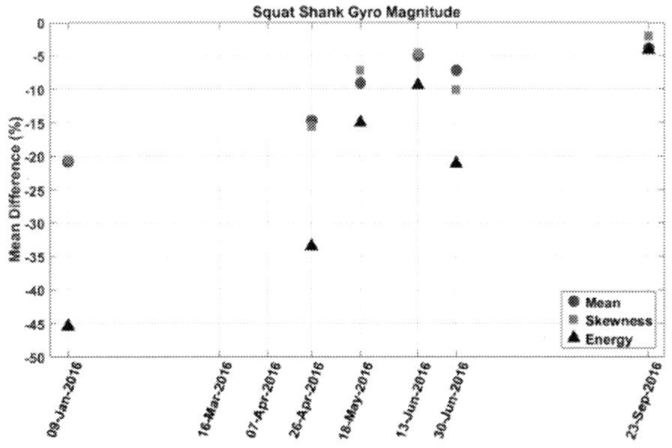

Figure 7. Mean difference for different statistical parameters in the half squat scenario showing the trends of progress during rehabilitation. Mean/skewness/energy for the gyro shank magnitude are shown

Similar considerations are evident on the gait test (Figures 8-10). The level of thigh movement, in this scenario, is not particularly informative, except for a reduced number of variables. Examples of these parameters are the root mean square calculated on the gyro magnitude and the standard deviation of the gyro around the vertical axis (when walking at 3 km/h), the CV of the acceleration/angular rate around the z-axis, and the maximum of the angular rate around the mediolateral axis (when walking at 4 km/h), and mean/peak-to-peak amplitude over the gyro x-axis, and the minimum of the gyro z-axis (when walking at 6 km/h). A higher number of helpful metrics is instead detected in the inertial data collected on the shank consistently for all the speeds. For instance, variables obtained from the acceleration magnitude can reliably show patient's progress over the rehabilitation course still highlighting the difference between left and right leg. Some parameters can be the mean, standard deviation, variance, level of skew, signal magnitude area, and energy when walking at slow speed, while increasing the speed may reduce the number of metrics to the standard deviation and

TABLE I. GAIT METRICS (3 KM/H)

GAIT 3KM/H	1st SESSION (09/01/2016)			2nd SESSION (16/03/16)			3rd SESSION (07/04/2016)			4th SESSION (26/04/16)		
Parameters	R	L	Mean Diff. (%) / Paerson's r / p-value	R	L	Mean Diff. (%) / Paerson's r / p-value	R	L	Mean Diff. (%) / Paerson's r / p-value	R	L	Mean Diff. (%) / Paerson's r / p-value
GCT (s)	1.386±0.030	1.346±0.102	-2.818/0.148/**0.045**	1.347±0.042	1.342±0.057	-0.397/0.356/0.671	1.387±0.032	1.379±0.030	-0.541/0.482/0.334	1.409±0.026	1.398±0.023	-0.755/0.457/0.090
CV_GCT (%)	2.177	7.574	247.880/-/-	3.107	4.27	37.414/-/-	2.309	2.152	-6.816/-/-	1.878	1.631	-13.155/-/-
B.I./Symmetry_GCT	0.022±0.039 / -3.895		-/-/-	0.015±0.015 / -0.863		-/-/-	0.008±0.009 / 0.113		-/-/-	0.007±0.007 / 0.175		-/-/-
Stance Phase (s)	0.886±0.026	0.864±0.020	-2.473/-0.042/**<0.001**	0.869±0.029	0.863±0.042	-0.707/0.101/0.500	0.896±0.026	0.871±0.018	-2.805/-0.057/**<0.001**	0.916±0.021	0.886±0.023	-3.232/0.211/**<0.001**
CV_Stance Phase (%)	2.945	2.34	-20.552/-/-	3.294	4.92	49.378/-/-	2.912	2.098	-27.965/-/-	2.287	2.609	14.116/-/-
B.I./Symmetry Stance Phase	0.019±0.013 / 3.887		-/-/-	0.024±0.014 / -0.485		-/-/-	0.019±0.013 / 0.039		-/-/-	0.018±0.014 / -0.700		-/-/-
Swing Phase (s)	0.500±0.011	0.483±0.103	-3.430/-0.089/0.356	0.478±0.016	0.479±0.051	0.165/0.339/0.933	0.491±0.019	0.509±0.028	3.589/0.167/**0.005**	0.493±0.013	0.512±0.010	3.842/0.016/**<0.001**
CV_Swing Phase (%)	2259	21.334	844.264/-/-	3.354	10.576	215.319/-/-	3.807	5.544	45.650/-/-	2.537	1.949	-23.182/-/-
B.I./Symmetry Swing Phase	0.063±0.142 / -20.762		-/-/-	0.044±0.026 / -8.383		-/-/-	0.030±0.021 / -2.938		-/-/-	0.020±0.015 / 0.369		-/-/-
Double Support (s)	0.403±0.101		-/-/-	0.390±0.050		-/-/-	0.389±0.028		-/-/-	0.403±0.019		-/-/-
CV_Double Support (%)	25.05		-/-/-	12.748		-/-/-	7.191		-/-/-	4.811		-/-/-
Stride Length (m)	1.179±0.068	0.895±0.068	-24.112/0.273/**<0.001**	1.252±0.089	0.932±0.132	-25.529/0.284/**<0.001**	1.104±0.105	0.972±0.081	-11.941/0.206/**<0.001**	1.173±0.148	0.938±0.093	-20.072/0.309/**<0.001**
CV_Stride Length (%)	58	7.56	30.350/-/-	7.084	14.119	99.324/-/-	9.509	8.316	-12.544/-/-	12.601	9.865	-21.710/-/-
B.I./Symmetry Stride Length	0.137±0.040 / -6.044		-/-/-	0.149±0.069 / -10.664		-/-/-	0.067±0.051 / -7.140		-/-/-	0.123±0.056 / -8.190		-/-/-
Stride Speed (m/s)	0.851±0.050	0.669±0.077	-21.434/-0.104/**<0.001**	0.931±0.078	0.695±0.097	-25.314/0.259/**<0.001**	0.796±0.073	0.704±0.052	-11.503/0.076/**<0.001**	0.833±0.103	0.671±0.064	-19.453/0.219/**<0.001**
CV_Stride Speed (%)	5831	11.518	97.526/-/-	8.327	13.894	66.841/-/-	9.159	7.378	-19.449/-/-	12.409	9.611	-22.549/-/-
B.I./Symmetry Stride Speed	0.126±0.053 / -18.524		-/-/-	0.147±0.072 / -22.033		-/-/-	0.065±0.050 / -7.272		-/-/-	0.121±0.059 / -14.342		-/-/-
Clearance (m)	0.034±0.008	0.013±0.005	-61.914/0.138/**<0.001**	0.060±0.016	0.023±0.006	-61.380/0.186/**<0.001**	0.036±0.012	0.022±0.006	-40.290/0.341/**<0.001**	0.030±0.009	0.019±0.005	-37.929/0.229/**<0.001**
CV_Clearance (%)	22889	41.738	82.348/-/-	26.463	26.497	0.130/-/-	32.459	25.341	-21.930/-/-	28.904	27.332	-5.438/-/-
B.I./Symmetry Clearance	0.453±0.166 / -42.273		-/-/-	0.432±0.144 / -33.229		-/-/-	0.254±0.142 / -38.919		-/-/-	0.241±0.148 / -49.557		-/-/-
Relative Stance Phase (%)	63.918±0.782	64.597±6.306	1.063/-0.246/0.550	64.489±0.550	64.337±2.966	-0.236/0.099/0.777	64.596±1.088	63.146±1.541	-2.246/-0.184/**<0.001**	64.986±0.695	63.359±0.840	-2.504/-0.220/**<0.001**
CV_Relative Stance Phase (%)	1223	9.761	697.930/-/-	0.852	4.61	441.018/-/-	1.685	2.44	44.828/-/-	1.07	1.326	23.977/-/-
B.I./Symmetry Relative St. Ph.	0.023±0.039 / 37.935		-/-/-	0.019±0.012 / 20.887		-/-/-	0.017±0.009 / 12.210		-/-/-	0.013±0.009 / 8.639		-/-/-
	5th SESSION (18/05/16)			6th SESSION (13/06/16)			7th SESSION (30/06/16)			8th SESSION (23/09/16)		
Parameters	R	L	Mean Diff. (%) / Paerson's r / p-value	R	L	Mean Diff. (%) / Paerson's r / p-value	R	L	Mean Diff. (%) / Paerson's r / p-value	R	L	Mean Diff. (%) / Paerson's r / p-value
GCT (s)	1.387±0.036	1.383±0.027	-0.272/0.088/0.637	1.464±0.049	1.455±0.048	-0.602/0.799/0.471	1.396±0.032	1.390±0.030	-0.385/0.667/0.495	1.365±0.054	1.360±0.047	-0.315/0.886/0.736
CV_GCT (%)	2.618	1.922	-26.577/-/-	3.343	3.316	-0.800/-/-	2.317	2.169	-6.386/-/-	3.96	3.467	-12.448/-/-
B.I./Symmetry_GCT	0.012±0.009 / 0.502		-/-/-	0.008±0.008 / 0.018		-/-/-	0.006±0.007 / 0.106		-/-/-	0.006±0.007 / 0.361		-/-/-
Stance Phase (s)	0.909±0.029	0.878±0.023	-3.380/-0.080/**<0.001**	0.964±0.048	0.936±0.044	-2.845/0.755/**0.021**	0.920±0.024	0.882±0.025	-4.079/0.447/**<0.001**	0.890±0.042	0.866±0.029	-2.705/0.691/**0.010**
CV_Stance Phase (%)	3.209	2.596	-19.102/-/-	5.011	4.689	-6.423/-/-	2.612	2.864	9.676/-/-	4.754	3.307	-30.437/-/-
B.I./Symmetry Stance Phase	0.022±0.016 / -0.487		-/-/-	0.016±0.015 / -0.943		-/-/-	0.021±0.014 / -0.500		-/-/-	0.015±0.015 / 0.117		-/-/-
Swing Phase (s)	0.479±0.015	0.505±0.019	5.629/0.353/**<0.001**	0.500±0.011	0.519±0.011	3.721/0.220/**<0.001**	0.476±0.016	0.508±0.015	6.753/0.305/**<0.001**	0.474±0.017	0.494±0.025	4.169/0.643/**<0.001**
CV_Swing Phase (%)	3.03	3.694	21.888/-/-	2.229	2.2	-1.276/-/-	3.393	2.955	-12.925/-/-	3.51	4.96	41.304/-/-
B.I./Symmetry Swing Phase	0.028±0.017 / -0.533		-/-/-	0.018±0.014 / 2.916		-/-/-	0.033±0.018 / -0.373		-/-/-	0.022±0.017 / -0.231		-/-/-
Double Support (s)	0.405±0.026		-/-/-	0.442±0.039		-/-/-	0.413±0.021		-/-/-	0.398±0.024		-/-/-
CV_Double Support (%)	6.361		-/-/-	8.796		-/-/-	5.063		-/-/-	6.156		-/-/-
Stride Length (m)	1.054±0.149	0.994±0.113	-5.708/0.481/0.073	1.276±0.053	1.071±0.052	-16.035/0.412/**<0.001**	1.035±0.133	0.971±0.061	-6.174/0.672/**0.017**	1.043±0.130	0.947±0.082	-9.197/0.626/**0.001**
CV_Stride Length (%)	14.1	11.348	-19.513/-/-	4.182	4.816	15.152/-/-	12.833	6.274	-51.111/-/-	12.472	8.659	-30.573/-/-
B.I./Symmetry Stride Length	0.058±0.044 / -9.969		-/-/-	0.087±0.025 / -0.136		-/-/-	0.041±0.036 / -3.865		-/-/-	0.053±0.040 / -6.179		-/-/-
Stride Speed (m/s)	0.759±0.101	0.718±0.076	-5.419/0.450/0.072	0.872±0.043	0.737±0.042	-15.516/0.432/**<0.001**	0.741±0.086	0.698±0.038	-5.741/0.582/**0.014**	0.763±0.077	0.696±0.048	-8.831/0.380/**<0.001**
CV_Stride Speed (%)	13.364	10.643	-20.355/-/-	4.928	5.693	15.516/-/-	11.635	5.454	-53.128/-/-	10.069	6.95	-30.976/-/-
B.I./Symmetry Stride Speed	0.054±0.044 / -15.909		-/-/-	0.084±0.029 / -6.925		-/-/-	0.039±0.036 / -4.329		-/-/-	0.052±0.041 / -7.249		-/-/-
Clearance (m)	0.034±0.006	0.033±0.009	-0.234/-0.126/0.968	0.032±0.007	0.017±0.004	-46.935/0.091/**<0.001**	0.033±0.006	0.018±0.004	-45.135/0.227/**<0.001**	0.031±0.006	0.026±0.010	-13.621/-0.457/**0.046**
CV_Clearance (%)	18.107	26.931	48.738/-/-	22.225	23.492	5.702/-/-	16.748	23.785	42.016/-/-	19.92	36.949	85.485/-/-
B.I./Symmetry Clearance	0.139±0.127 / -45.973		-/-/-	0.303±0.139 / -41.042		-/-/-	0.294±0.115 / -41.001		-/-/-	0.212±0.156 / -64.720		-/-/-
Relative Stance Phase (%)	65.497±0.830	63.463±1.147	-3.106/0.125/**<0.001**	65.805±1.189	64.325±1.066	-2.250/0.524/**<0.001**	65.629±1.817	63.285±0.975	-3.715/-0.033/**<0.001**	65.238±0.792	63.706±0.894	-2.357/-0.307/**<0.001**
CV_Relative Stance Phase (%)	1.268	1.807	42.553/-/-	1.807	1.657	-8.332/-/-	1.249	1.459	16.879/-/-	1.325	1.475	11.373/-/-
B.I./Symmetry Relative St. Ph.	0.016±0.009 / 11.253		-/-/-	0.011±0.008 / 16.152		-/-/-	0.019±0.010 / 8.737		-/-/-	0.013±0.010 / 11.764		-/-/-

TABLE II. GAIT METRICS (4 KM/H)

GAIT 4KM/H	1st SESSION (09/01/2016)			2nd SESSION (16/03/16)			3rd SESSION (07/04/2016)			4th SESSION (26/04/16)		
Parameters	R	L	Mean Diff. (%) / Paerson's r / p-value	R	L	Mean Diff. (%) / Paerson's r / p-value	R	L	Mean Diff. (%) / Paerson's r / p-value	R	L	Mean Diff. (%) / Paerson's r / p-value
GCT (s)	1.166±0.024	1.166±0.021	-0.038/0.739/0.943	1.186±0.018	1.180±0.064	-0.479/0.049/0.662	1.216±0.018	1.217±0.016	0.055/0.670/0.879	1.131±0.025	1.131±0.025	0.075/0.939/0.897
CV_GCT (%)	2.097	1.818	-13.271/-/-	1.476	5.465	270.189/-/-	1.441	1.314	-8.842/-/-	2.213	2.242	1.34/-/-
B.I./Symmetry_GCT	0.005±0.005 / 0.239		-/-/-	0.018±0.022 / -3.364		-/-/-	0.004±0.004 / 0.105		-/-/-	0.003±0.002 / -0.026		-/-/-
Stance Phase (s)	0.718±0.020	0.709±0.015	-1.318/0.659/0.056	0.734±0.017	0.770±0.055	4.842/0.374/**0.003**	0.765±0.015	0.745±0.014	-2.686/0.659/**<0.001**	0.705±0.019	0.691±0.018	-2.000/0.814/**0.005**
CV_Stance Phase (%)	2.787	2.141	-23.183/-/-	2.321	7.189	209.778/-/-	2.023	1.828	-9.644/-/-	2.745	2.619	-4.594/-/-
B.I./Symmetry Stance Phase	0.010±0.007 / -0.277		-/-/-	0.033±0.022 / -1.461		-/-/-	0.014±0.007 / -0.423		-/-/-	0.011±0.007 / -0.332		-/-/-
Swing Phase (s)	0.448±0.011	0.457±0.012	2.014/0.429/**0.005**	0.451±0.010	0.410±0.063	-9.141/0.079/**0.002**	0.451±0.009	0.472±0.006	4.708/0.042/**<0.001**	0.426±0.009	0.441±0.009	3.513/0.537/**<0.001**
CV_Swing Phase (%)	2.42	2.604	7.587/-/-	2.319	15.33	561.136/-/-	2.099	1.244	-40.724/-/-	2.169	2.096	-3.359/-/-
B.I./Symmetry Swing Phase	0.014±0.008 / 0.255		-/-/-	0.070±0.066 / -17.715		-/-/-	0.024±0.009 / 1.017		-/-/-	0.018±0.010 / 0.920		-/-/-
Double Support (s)	0.262±0.020		-/-/-	0.320±0.057		-/-/-	0.293±0.016		-/-/-	0.265±0.012		-/-/-
CV_Double Support (%)	7.635		-/-/-	17.798		-/-/-	5.382		-/-/-	4.465		-/-/-
Stride Length (m)	1.298±0.070	1.111±0.063	-14.431/0.532/**<0.001**	1.353±0.057	1.088±0.082	-19.603/0.031/**<0.001**	1.202±0.091	1.058±0.046	-11.932/0.117/**<0.001**	1.171±0.045	0.785±0.163	-32.901/0.478/**<0.001**
CV_Stride Length (%)	5.371	5.626	4.751/-/-	4.185	7.524	79.773/-/-	7.531	4.37	-41.977/-/-	3.843	20.766	440.346/-/-
B.I./Symmetry Stride Length	0.078±0.026 / -4.916		-/-/-	0.109±0.042 / -0.436		-/-/-	0.065±0.036 / -3.413		-/-/-	0.206±0.100 / -26.265		-/-/-
Stride Speed (m/s)	1.114±0.065	0.953±0.044	-14.462/0.388/**<0.001**	1.141±0.045	0.923±0.073	-19.080/-0.043/**<0.001**	0.989±0.077	0.870±0.041	-11.984/0.195/**<0.001**	1.036±0.046	0.693±0.136	-33.120/0.226/**<0.001**
CV_Stride Speed (%)	5.842	4.608	-21.134/-/-	3.936	7.903	100.776/-/-	7.824	4.713	-39.754/-/-	4.416	19.66	345.173/-/-
B.I./Symmetry Stride Speed	0.078±0.028 / -4.885		-/-/-	0.106±0.044 / -12.376		-/-/-	0.066±0.036 / -5.800		-/-/-	0.206±0.100 / -41.100		-/-/-
Clearance (m)	0.048±0.014	0.017±0.006	-63.693/0.062/**<0.001**	0.057±0.010	0.025±0.007	-57.109/-0.166/**<0.001**	0.049±0.014	0.026±0.003	-48.108/-0.080/**<0.001**	0.042±0.006	0.025±0.003	-40.303/0.587/**<0.001**
CV_Clearance (%)	28.276	34.606	22.383/-/-	17.093	29.702	73.763/-/-	27.537	13.102	-52.42/-/-	14.489	13.589	-6.211/-/-
B.I./Symmetry Clearance	0.462±0.154 / -70.054		-/-/-	0.402±0.153 / -35.059		-/-/-	0.299±0.137 / -25.127		-/-/-	0.251±0.061 / -26.089		-/-/-
Relative Stance Phase (%)	61.596±0.793	60.813±0.717	-1.271/0.285/**<0.001**	61.940±0.873	65.335±4.507	5.482/0.284/**0.001**	62.932±0.711	61.207±0.469	-2.740/0.204/**<0.001**	62.354±0.603	61.062±0.456	-2.072/-0.110/**<0.001**
CV_Relative Stance Phase (%)	1.287	1.179	-8.439/-/-	1.409	6.898	389.572/-/-	1.13	0.766	-32.259/-/-	0.967	0.747	-22.797/-/-
B.I./Symmetry Relative St. Ph.	0.008±0.005 / 24.685		-/-/-	0.032±0.026 / 34.111		-/-/-	0.014±0.005 / 15.737		-/-/-	0.011±0.006 / 14.024		-/-/-

	5th SESSION (18/05/16)			6th SESSION (13/06/16)			7th SESSION (30/06/16)			8th SESSION (23/09/16)		
Parameters	R	L	Mean Diff. (%) / Paerson's r / p-value	R	L	Mean Diff. (%) / Paerson's r / p-value	R	L	Mean Diff. (%) / Paerson's r / p-value	R	L	Mean Diff. (%) / Paerson's r / p-value
GCT (s)	1.154±0.017	1.156±0.015	0.194/0.480/0.574	1.292±0.107	1.292±0.109	0.037/0.869/0.986	1.088±0.020	1.088±0.028	0.003/0.659/0.995	1.101±0.022	1.101±0.022	-0.037/0.779/0.937
CV_GCT (%)	1.48	1.264	-14.567/-/-	8.308	8.406	1.183/-/-	1.86	2.596	39.604/-/-	1.969	2.038	3.537/-/-
B.I./Symmetry_GCT	0.005±0.005 / 0.187		-/-/-	0.013±0.016 / -0.076		-/-/-	0.006±0.008 / -0.677		-/-/-	0.005±0.004 / -0.063		-/-/-
Stance Phase (s)	0.726±0.013	0.709±0.013	-2.243/0.202/**<0.001**	0.837±0.087	0.803±0.079	-4.097/0.848/0.102	0.684±0.016	0.660±0.021	-3.539/0.510/**<0.001**	0.689±0.025	0.673±0.018	-2.312/0.653/**0.003**
CV_Stance Phase (%)	1.729	1.81	4.636/-/-	10.33	9.807	-5.06/-/-	2.331	3.174	36.197/-/-	3.691	2.603	-29.462/-/-
B.I./Symmetry Stance Phase	0.013±0.009 / -0.472		-/-/-	0.026±0.022 / -1.084		-/-/-	0.018±0.014 / -0.531		-/-/-	0.015±0.010 / -0.513		-/-/-
Swing Phase (s)	0.428±0.009	0.447±0.006	4.327/0.174/**<0.001**	0.454±0.033	0.489±0.036	7.655/0.377/**<0.001**	0.404±0.006	0.429±0.018	5.993/0.238/**<0.001**	0.412±0.018	0.428±0.008	3.766/0.222/**<0.001**
CV_Swing Phase (%)	2.18	1.407	-35.459/-/-	7.199	7.42	3.059/-/-	1.606	4.247	164.401/-/-	4.264	1.754	-58.853/-/-
B.I./Symmetry Swing Phase	0.021±0.012 / 0.445		-/-/-	0.043±0.029 / 3.476		-/-/-	0.029±0.020 / -2.801		-/-/-	0.023±0.014 / 2.810		-/-/-
Double Support (s)	0.281±0.011		-/-/-	0.349±0.068		-/-/-	0.255±0.020		-/-/-	0.262±0.024		-/-/-
CV_Double Support (%)	3.821		-/-/-	19.603		-/-/-	7.845		-/-/-	9.329		-/-/-
Stride Length (m)	1.178±0.085	1.156±0.076	-1.887/0.303/0.277	1.162±0.105	0.979±0.147	-15.759/-0.192/**<0.001**	0.989±0.047	0.780±0.156	-21.188/0.076/**<0.001**	1.088±0.118	0.952±0.092	-12.496/0.104/**<0.001**
CV_Stride Length (%)	7.255	6.618	-8.78/-/-	9.05	15.056	66.358/-/-	4.744	20.014	321.893/-/-	10.867	9.633	-11.363/-/-
B.I./Symmetry Stride Length	0.032±0.031 / -6.780		-/-/-	0.108±0.066 / -6.536		-/-/-	0.129±0.117 / -25.524		-/-/-	0.084±0.051 / -10.441		-/-/-
Stride Speed (m/s)	1.021±0.068	1.000±0.063	-2.055/0.228/0.206	0.904±0.088	0.768±0.166	-14.967/0.341/**<0.001**	0.910±0.052	0.717±0.145	-21.199/0.143/**<0.001**	0.988±0.101	0.865±0.076	-12.469/-0.116/**<0.001**
CV_Stride Speed (%)	6.633	6.344	-4.363/-/-	9.773	21.578	120.794/-/-	5.675	20.184	255.634/-/-	10.199	8.785	-13.865/-/-
B.I./Symmetry Stride Speed	0.031±0.031 / -9.729		-/-/-	0.107±0.071 / -31.646		-/-/-	0.129±0.117 / -45.932		-/-/-	0.085±0.053 / -10.971		-/-/-
Clearance (m)	0.047±0.008	0.041±0.005	-11.899/-0.405/**0.002**	0.039±0.009	0.030±0.010	-23.242/0.027/**<0.001**	0.040±0.006	0.027±0.009	-34.109/-0.156/**<0.001**	0.034±0.011	0.035±0.010	2.572/-0.404/0.722
CV_Clearance (%)	17.009	12.597	-25.94/-/-	23.676	33.781	42.676/-/-	14.293	35.24	146.549/-/-	31.821	28.428	-10.663/-/-
B.I./Symmetry Clearance	0.104±0.088 / -25.062		-/-/-	0.186±0.135 / -53.894		-/-/-	0.244±0.133 / -72.296		-/-/-	0.207±0.112 / -62.369		-/-/-
Relative Stance Phase (%)	62.898±0.582	61.365±0.529	-2.438/-0.244/**<0.001**	63.303±0.640	60.770±0.635	-4.088/0.172/**<0.001**	62.835±0.462	60.615±1.225	-3.533/-0.098/**<0.001**	62.569±1.641	61.148±0.529	-2.271/0.382/**<0.001**
CV_Relative Stance Phase (%)	0.926	0.862	-6.896/-/-	3.054	2.406	-21.242/-/-	0.735	2.021	174.96/-/-	2.622	0.865	-67.002/-/-
B.I./Symmetry Relative St. Ph.	0.012±0.007 / 10.518		-/-/-	0.023±0.016 / 49.315		-/-/-	0.018±0.011 / 22.814		-/-/-	0.014±0.009 / 32.365		-/-/-

TABLE III. GAIT METRICS (6 KM/H)

GAIT 6KM/H	1st SESSION (09/01/2016)			4th SESSION (26/04/16)			5th SESSION (18/05/16)		
Parameters	R	L	Mean Diff. (%) / Paerson's r / p-value	R	L	Mean Diff. (%) / Paerson's r / p-value	R	L	Mean Diff. (%) / Paerson's r / p-value
GCT (s)	0.925±0.016	0.925±0.017	-0.032/-0.148/0.941	0.955±0.010	0.954±0.012	-0.021/0.517/0.935	0.920±0.010	0.920±0.013	0.007/0.708/0.980
CV_GCT (%)	1.747	1.813	3.798/-/-	1.059	1.247	17.766/-/-	1.11	1.449	30.442/-/-
B.I./Symmetry_GCT	0.007±0.011 / -0.072		-/-/-	0.005±0.003 / -0.197		-/-/-	0.004±0.003 / -0.367		-/-/-
Stance Phase (s)	0.546±0.010	0.532±0.013	-2.573/-0.109/**<0.001**	0.571±0.008	0.559±0.010	-2.089/0.347/**<0.001**	0.543±0.010	0.533±0.011	-1.884/0.615/**<0.001**
CV_Stance Phase (%)	1.9	2.438	28.303/-/-	1.381	1.701	23.159/-/-	1.797	2.08	15.741/-/-
B.I./Symmetry Stance Phase	0.016±0.014 / -0.675		-/-/-	0.012±0.007 / -0.476		-/-/-	0.011±0.006 / -0.686		-/-/-
Swing Phase (s)	0.380±0.012	0.393±0.013	3.619/0.043/**<0.001**	0.383±0.006	0.395±0.006	3.060/0.269/**<0.001**	0.377±0.005	0.387±0.006	2.731/0.055/**<0.001**
CV_Swing Phase (%)	3.044	3.345	9.914/-/-	1.527	1.611	5.542/-/-	1.216	1.465	20.494/-/-
B.I./Symmetry Swing Phase	0.023±0.015 / -2.649		-/-/-	0.016±0.009 / -0.402		-/-/-	0.015±0.007 / 0.610		-/-/-
Double Support (s)	0.152±0.021		-/-/-	0.176±0.010		-/-/-	0.156±0.011		-/-/-
CV_Double Support (%)	13.648		-/-/-	5.851		-/-/-	7.189		-/-/-
Stride Length (m)	1.346±0.090	1.151±0.190	-14.526/0.280/**<0.001**	1.358±0.042	1.152±0.082	-15.112/0.181/**<0.001**	1.310±0.046	1.232±0.063	-5.954/-0.144/**<0.001**
CV_Stride Length (%)	6.683	16.509	147.047/-/-	3.083	7.159	132.205/-/-	3.501	5.073	44.911/-/-
B.I./Symmetry Stride Length	0.086±0.098 / -26.474		-/-/-	0.083±0.037 / -9.758		-/-/-	0.037±0.025 / -5.618		-/-/-
Stride Speed (m/s)	1.456±0.105	1.244±0.208	-14.519/0.170/**<0.001**	1.422±0.048	1.208±0.090	-15.082/0.288/**<0.001**	1.424±0.048	1.339±0.065	-5.963/-0.283/**<0.001**
CV_Stride Speed (%)	7.238	16.686	130.552/-/-	3.368	7.424	120.39/-/-	3.353	4.829	44.027/-/-
B.I./Symmetry Stride Speed	0.089±0.098 / -35.941		-/-/-	0.083±0.037 / -15.383		-/-/-	0.038±0.024 / -9.582		-/-/-
Clearance (m)	0.066±0.014	0.032±0.011	-51.184/0.453/**<0.001**	0.057±0.009	0.033±0.006	-42.273/-0.115/**<0.001**	0.058±0.007	0.047±0.008	-18.035/-0.027/**<0.001**
CV_Clearance (%)	20.361	35.015	71.97/-/-	15.214	19.4	27.513/-/-	11.6	16.631	43.362/-/-
B.I./Symmetry Clearance	0.354±0.130 / -80.517		-/-/-	0.268±0.115 / -45.027		-/-/-	0.115±0.090 / -39.151		-/-/-
Relative Stance Phase (%)	58.973±0.821	57.473±1.093	-2.544/0.105/**<0.001**	59.841±0.481	58.603±0.541	-2.070/0.065/**<0.001**	59.018±0.559	57.902±0.566	-1.892/0.193/**<0.001**
CV_Relative Stance Phase (%)	1.393	1.902	36.54/-/-	0.804	0.922	14.776/-/-	0.947	0.978	3.177/-/-
B.I./Symmetry Relative St. Ph.	0.015±0.009 / 77.317		-/-/-	0.011±0.006 / 28.012		-/-/-	0.010±0.005 / 39.871		-/-/-

	6th SESSION (13/06/16)			7th SESSION (30/06/16)			8th SESSION (23/09/16)		
Parameters	R	L	Mean Diff. (%) / Paerson's r / p-value	R	L	Mean Diff. (%) / Paerson's r / p-value	R	L	Mean Diff. (%) / Paerson's r / p-value
GCT (s)	0.900±0.008	0.900±0.011	-0.003/0.528/0.989	0.886±0.010	0.887±0.028	0.037/0.335/0.944	0.901±0.061	0.925±0.030	2.617/0.348/**0.029**
CV_GCT (%)	0.917	1.169	27.422/-/-	1.098	3.192	190.574/-/-	6.766	3.296	-51.281/-/-
B.I./Symmetry_GCT	0.004±0.003 / -0.279		-/-/-	0.008±0.013 / -2.362		-/-/-	0.018±0.032 / 3.850		-/-/-
Stance Phase (s)	0.535±0.008	0.518±0.010	-3.276/0.233/**<0.001**	0.524±0.007	0.509±0.022	-2.982/0.261/**<0.001**	0.549±0.019	0.538±0.021	-1.914/0.856/**0.021**
CV_Stance Phase (%)	1.51	1.938	28.318/-/-	1.415	4.395	210.523/-/-	3.536	3.956	11.871/-/-
B.I./Symmetry Stance Phase	0.017±0.010 / -0.855		-/-/-	0.017±0.021 / -1.357		-/-/-	0.012±0.008 / -0.713		-/-/-
Swing Phase (s)	0.365±0.007	0.383±0.006	4.796/0.020/**<0.001**	0.362±0.006	0.378±0.027	4.407/0.081/**<0.001**	0.352±0.057	0.387±0.012	9.670/0.041/**<0.001**
CV_Swing Phase (%)	1.806	1.653	-8.436/-/-	1.576	7.054	347.579/-/-	16.272	3.169	-80.525/-/-
B.I./Symmetry Swing Phase	0.024±0.010 / -0.267		-/-/-	0.022±0.031 / -10.758		-/-/-	0.054±0.099 / 0.669		-/-/-
Double Support (s)	0.153±0.010		-/-/-	0.146±0.023		-/-/-	0.163±0.013		-/-/-
CV_Double Support (%)	6.848		-/-/-	15.677		-/-/-	7.775		-/-/-
Stride Length (m)	1.301±0.030	1.202±0.050	-7.566/0.465/**<0.001**	1.211±0.051	0.786±0.165	-35.136/0.176/**<0.001**	1.261±0.060	0.898±0.130	-28.778/0.414/**<0.001**
CV_Stride Length (%)	2.341	4.161	77.771/-/-	4.233	20.979	395.61/-/-	4.734	14.437	204.957/-/-
B.I./Symmetry Stride Length	0.040±0.019 / -4.293		-/-/-	0.221±0.098 / -32.612		-/-/-	0.172±0.067 / -19.476		-/-/-
Stride Speed (m/s)	1.445±0.037	1.336±0.055	-7.567/0.478/**<0.001**	1.367±0.056	0.886±0.184	-35.144/0.085/**<0.001**	1.406±0.119	0.970±0.132	-30.982/0.108/**<0.001**
CV_Stride Speed (%)	2.54	4.108	61.72/-/-	4.064	20.764	410.956/-/-	8.435	13.582	61.019/-/-
B.I./Symmetry Stride Speed	0.040±0.018 / -6.307		-/-/-	0.221±0.099 / -52.995		-/-/-	0.185±0.073 / 7.632		-/-/-
Clearance (m)	0.058±0.006	0.041±0.008	-29.571/0.120/**<0.001**	0.048±0.004	0.037±0.013	-24.058/0.011/**<0.001**	0.045±0.006	0.058±0.014	28.036/-0.204/**<0.001**
CV_Clearance (%)	10.61	19.632	85.033/-/-	9.04	35.723	295.162/-/-	13.914	24.693	77.468/-/-
B.I./Symmetry Clearance	0.181±0.095 / -46.997		-/-/-	0.184±0.122 / -75.838		-/-/-	0.178±0.103 / -55.680		-/-/-
Relative Stance Phase (%)	59.456±0.663	57.508±0.692	-3.276/-0.073/**<0.001**	59.145±0.489	57.380±2.216	-2.990/0.038/**<0.001**	61.147±4.604	58.188±0.761	-4.839/0.017/**<0.001**
CV_Relative Stance Phase (%)	1.116	1.203	7.823/-/-	0.826	3.861	367.397/-/-	7.53	1.309	-82.621/-/-
B.I./Symmetry Relative St. Ph.	0.017±0.008 / 36.958		-/-/-	0.016±0.020 / 80.676		-/-/-	0.024±0.035 / 39.576		-/-/-

TABLE IV. KINEMATICS METRICS - RANGE OF MOTION

	Parameters	1st SESSION (09/01/2016) R	L	Mean Diff. (%)	2nd SESSION (16/03/16) R	L	Mean Diff. (%)	3rd SESSION (07/04/2016) R	L	Mean Diff. (%)	4th SESSION (26/04/16) R	L	Mean Diff. (%)
Hamstring Curl	ROM X-axis	-	82.641±2.737	-	118.172±1.482	67.951±7.716	-73.908	111.836±3.241	91.923±1.286	-21.662	112.139±8.342	107.131±1.703	-4.674
	ROM Y-axis	-	15.442±1.372	-	9.575±1.286	9.020±1.303	-6.152	8.221±1.204	5.572±0.663	-47.551	9.983±2.843	8.373±0.962	-19.236
	ROM Z-axis	-	27.078±0.947	-	14.677±1.592	13.410±1.025	-9.445	13.657±1.141	9.230±1.027	-47.965	14.078±1.083	15.175±1.587	7.229
Flexion - Extension	ROM X-axis	-	-	-	122.970±1.671	100.828±0.665	-21.961	128.016±1.655	110.347±1.379	-16.012	130.550±1.313	52.832±7.771	-147.101
	ROM Y-axis	-	-	-	11.910±1.204	5.595±0.759	-112.887	26.833±1.859	18.286±1.322	-46.742	39.836±1.530	21.647±1.962	-84.02
	ROM Z-axis	-	-	-	13.209±1.298	5.724±0.264	-130.751	13.983±1.566	9.421±0.548	-48.429	13.371±1.222	10.190±0.951	-31.214
Squat	ROM X-axis	85.137±1.556	77.938±1.957	-9.236	-	-	-	-	-	-	93.786±23.775	84.844±20.745	-10.539
	ROM Y-axis	13.301±1.066	14.791±1.132	10.074	-	-	-	-	-	-	7.775±1.439	10.368±1.331	25.012
	ROM Z-axis	20.824±1.257	7.271±1.185	-186.408	-	-	-	-	-	-	20.032±4.139	13.505±1.464	-48.333
Gait - 3km/h	ROM X-axis	64.315±2.097	69.389±1.564	7.313	66.998±1.193	59.320±2.845	-12.944	61.618±1.524	60.458±1.856	-1.918	61.569±1.959	62.875±2.153	2.077
	ROM Y-axis	15.569±0.946	32.374±1.626	51.909	9.667±1.347	7.738±0.923	-24.937	14.965±1.142	7.379±0.790	-102.789	12.786±1.458	7.635±1.020	-67.455
	ROM Z-axis	37.565±101.720	42.273±96.859	11.137	13.050±1.545	10.121±1.170	-28.935	56.325±127.917	54.878±118.030	-2.637	61.721±121.070	44.364±106.139	-39.124
Gait - 4km/h	ROM X-axis	69.369±1.943	62.329±2.195	-11.294	74.203±1.353	70.244±1.834	-5.636	75.348±1.404	68.997±2.261	-9.205	66.626±1.461	63.836±2.433	-4.37
	ROM Y-axis	21.117±1.394	27.893±1.524	24.292	18.488±1.456	11.364±1.093	-62.685	13.404±1.347	9.797±1.206	-36.828	12.007±1.173	9.230±1.234	-30.097
	ROM Z-axis	43.085±111.755	16.996±1.488	-153.494	37.707±71.355	97.983±170.923	61.517	29.822±62.300	51.402±114.929	41.982	70.068±117.421	86.848±137.315	19.321
Gait - 6km/h	ROM X-axis	74.423±2.902	74.423±2.902	0	-	-	-	-	-	-	62.210±3.627	62.975±2.834	1.215
	ROM Y-axis	16.550±1.099	16.578±1.087	0.167	-	-	-	-	-	-	10.882±2.085	11.073±2.634	1.719
	ROM Z-axis	0.253±0.030	0.485±0.053	47.949	-	-	-	-	-	-	0.353±0.034	0.260±0.043	-35.844

	Parameters	5th SESSION (18/05/16) R	L	Mean Diff. (%)	6th SESSION (13/06/16) R	L	Mean Diff. (%)	7th SESSION (30/06/16) R	L	Mean Diff. (%)	8th SESSION (23/09/16) R	L	Mean Diff. (%)
Hamstring Curl	ROM X-axis	117.846±3.826	103.190±1.148	-14.203	122.307±1.865	97.093±3.927	-25.968	124.366±4.442	107.887±2.183	-15.275	108.661±3.817	104.718±2.630	-3.766
	ROM Y-axis	13.977±1.564	6.199±0.885	-125.463	16.278±1.389	22.596±2.363	27.964	12.218±1.921	5.345±0.996	-128.599	8.360±0.624	12.970±3.672	35.544
	ROM Z-axis	12.930±1.294	10.522±1.073	-22.887	8.931±1.843	12.545±1.539	28.806	11.295±1.444	10.343±1.556	-9.203	15.470±2.234	9.770±1.790	-58.332
Flexion - Extension	ROM X-axis	131.971±0.905	117.418±18.965	-12.393	131.493±0.937	116.473±20.045	-12.896	118.717±30.309	114.172±0.779	-3.981	129.839±1.810	125.422±2.211	-3.522
	ROM Y-axis	38.799±1.058	8.743±2.887	-343.755	52.276±0.934	14.720±1.769	-255.133	36.921±8.907	20.145±0.750	-83.279	29.963±1.160	14.866±2.571	-101.554
	ROM Z-axis	14.060±1.322	24.059±5.151	41.56	10.785±1.602	20.287±3.439	46.836	10.558±2.134	7.827±1.082	-34.894	22.439±1.579	50.844±5.072	55.867
Squat	ROM X-axis	101.020±1.566	95.359±1.633	-5.937	107.942±2.646	105.372±1.911	-2.439	112.705±1.005	104.399±1.400	-7.956	113.666±2.134	111.234±1.787	-2.187
	ROM Y-axis	6.575±0.659	11.905±1.118	44.773	11.268±1.075	15.684±0.878	28.156	7.939±0.644	11.326±0.903	29.911	11.578±0.362	13.253±0.599	12.641
	ROM Z-axis	21.238±0.862	9.852±1.292	-115.58	35.327±3.577	9.488±1.452	-272.322	24.207±0.927	6.238±0.587	-288.068	26.917±1.753	30.551±1.283	11.894
Gait - 3km/h	ROM X-axis	60.579±2.259	66.247±2.637	8.555	72.426±1.491	70.794±1.900	-2.306	55.414±2.689	51.661±1.485	-7.265	53.176±1.138	51.526±1.278	-3.202
	ROM Y-axis	10.054±0.959	5.911±0.841	-70.108	14.852±0.777	10.702±1.046	-38.774	8.995±1.346	9.393±0.985	4.244	15.110±1.223	11.166±0.719	-35.314
	ROM Z-axis	11.962±1.459	12.904±1.416	7.3	41.660±88.841	56.070±122.277	25.701	33.680±68.179	41.548±102.441	18.937	38.943±93.873	33.845±76.223	-15.063
Gait - 4km/h	ROM X-axis	63.176±1.906	64.103±1.491	1.445	70.553±1.306	70.143±2.085	-0.584	54.296±1.555	67.237±2.838	19.246	57.188±3.058	55.030±2.635	-3.922
	ROM Y-axis	10.539±0.977	7.519±0.575	-40.174	15.843±1.062	11.567±1.421	-36.964	10.335±0.930	13.904±1.465	25.668	15.116±1.227	12.386±1.086	-22.043
	ROM Z-axis	52.219±111.187	65.127±130.501	19.821	31.082±73.570	60.803±106.116	48.88	65.515±107.396	89.535±136.189	26.827	65.964±111.284	92.638±143.424	28.794
Gait - 6km/h	ROM X-axis	67.009±1.945	67.382±1.628	0.553	70.296±3.683	70.785±3.341	0.69	81.588±2.609	81.360±2.792	-0.281	67.352±4.163	67.285±4.439	-0.1
	ROM Y-axis	8.530±2.462	9.406±1.720	9.31	14.178±2.371	14.433±1.652	1.765	23.515±2.517	23.850±3.276	1.404	15.568±2.225	16.043±2.186	2.966
	ROM Z-axis	0.288±0.039	0.397±0.037	27.51	0.318±0.040	0.372±0.022	14.394	0.380±0.054	0.303±0.031	-25.552	0.389±0.043	0.363±0.045	-7.404

variance (at 4 km/h), and level of skew and energy (at 6 km/h). Analyzing the decomposed acceleration around the three components, other useful metrics are provided by the peak-to-peak amplitude/CV around the vertical axis (for both 3-4 km/h), standard deviation over the sagittal plane (at 3 km/h), mean around the x-axis (at 4 km/h), and the peak-to-peak amplitude over the x-axis, and the y-axis maximum (at 6 km/h). Finally, also the angular rate over its three components shows some trends. For example, the peak-to-peak amplitude over the sagittal axis is present at every speed, along with the y-axis standard deviation and PP at 4 km/h. The minimum over the x-axis is a parameter showing specific trends at every speed as well, together with the x-axis maximum/y-axis minimum at 3 km/h.

It is also evident how some of those features present a monotonic trend after the second session, showing a relevant performance gap between the first two sessions, due to the impact of the surgery event on the patient's mobility.

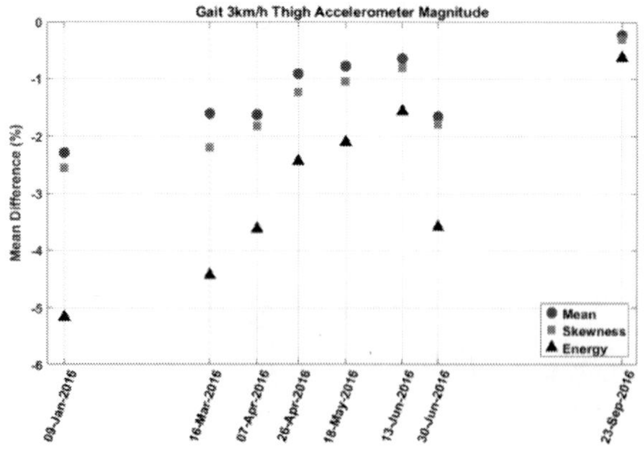

Figure 8. Mean difference for different statistical parameters in the gait scenario (3 km/h) showing the trends of progress during rehabilitation. Mean/skewness/energy for the acceleration thigh magnitude are shown

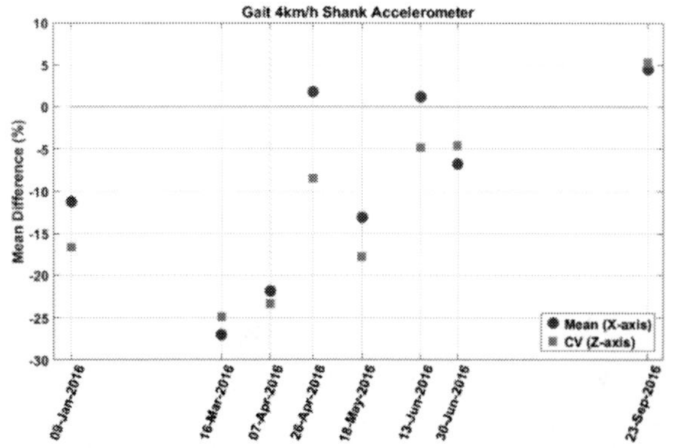

Figure 9. Mean difference for different statistical parameters in the gait scenario (4 km/h) showing the trends of progress during rehabilitation. Mean/CV for the acceleration shank x-, z-axis are shown

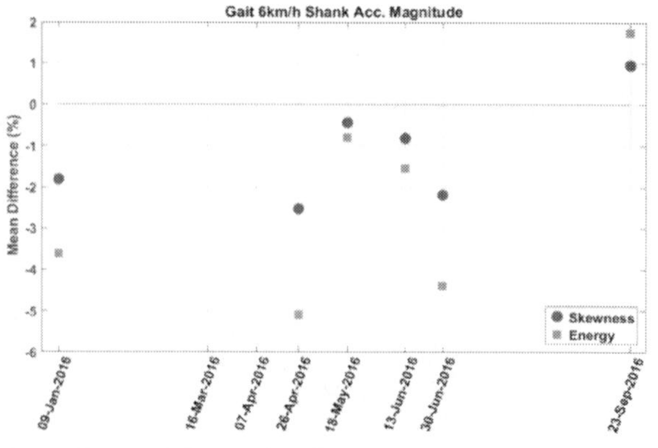

Figure 10. Mean difference for different statistical parameters in the gait scenario (6 km/h) showing the trends of progress during rehabilitation. Skewness/energy for the acceleration shank magnitude are shown

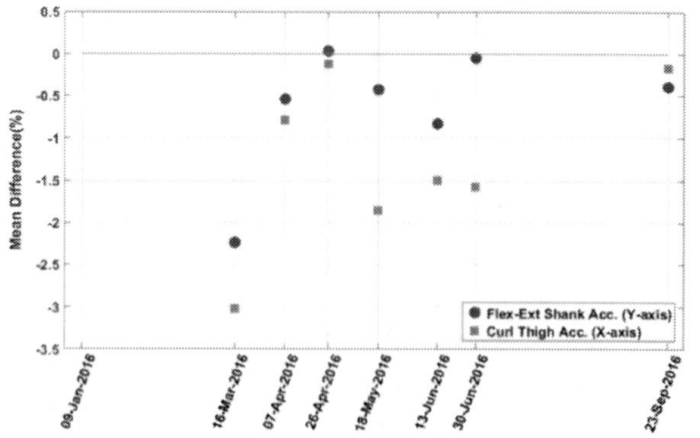

Figure 11. Mean difference for autocorrelation values in different scenarios showing the trends of progress during rehabilitation. X-axis thigh acceleration/y-axis shank acceleration for the hamstring curl/flexion-extension are shown

Finally, biased autocorrelation (e.g., the amplitude of the first dominant) has also been considered for all the scenarios (Figure 11). It has been calculated on both acceleration and angular rate from thigh and shank, but the computation has not been carried out on the single repetitions/strides but on the whole recorded session. In the hamstring curl test, autocorrelation on the x-axis of both acceleration and angular rate signals shows specific trends. On the other side, only y-axis autocorrelation for the shank acceleration was helpful in the flexion/extension scenario. Moreover, x-, z-axis autocorrelation in the shank acceleration and y-, z-axis autocorrelation in the thigh angular rate presented an observable trend in the half squat test. Eventually, in the walking scenario, the z-axis autocorrelation in the acceleration (at 3-4 km/h) was detected as for the thigh, while the x-axis autocorrelation for the gyro (4 km/h) and the y-axis autocorrelation in the acceleration (at 6 km/h) were noticed on the shank.

iii. Kinematics Features

Finally, the kinematic features described in Section IV have been calculated as well for each repetition/stride for all the four WIMUs adopted and for both acceleration and angular rate signals. Results are summarized for all the scenarios in the tables in the Appendix and discussed below. The table related to the ROM calculation is also integrated in the paper as an example (Table IV).

In the hamstring curl scenario, most of the detected features are present on the sagittal plane, given that the movement is mostly performed on this plane. The ROM over the x-axis is a clear example, while thigh and shank present similar results related to all the jerk-measures along the y-axis, and the fluency metrics over the sagittal plane. Moreover, the shank shows additional helpful parameters in the RANG and the vertical acceleration feature.

Similar results are shown for the flexion/extension scenario. X-axis ROM is one of the key metrics for highlighting patient progress, together with RANG shared by both thigh and shank. The thigh is also characterized by other metrics (in particular, jerk-measures N_MSJ, IAJ, and M_MAJ over both the x-, z-axis), while the metrics associated to the shank can be reduced to most of the jerk-measures and fluency both along the z-axis, as well as KV.

In the half squat test, again, ROM over the x-axis proved to be a helpful metrics, together with all the jerk-measures over the y-axis (z-axis) obtained from the thigh (shank). Additional metrics were detected in the z-axis stability for the thigh, and the RANG and vertical velocity for the shank. Results for these three scenarios are illustrated in Figure 12.

Gait tests have been also analyzed by using the kinematics variables. While ROM is not always showing particular trends (except at high speed over the x- and z-axis), there is a certain similarity among the feature detection at different speeds. For example, all jerk-measures over the anteroposterior axis on the thigh present clear convergence trends for every speed, while RANG and fluency over the thigh y-axis are more evident at 4-6 km/h, and stability is instead highlighted at slower speeds (3-4 km/h). Likewise, regarding the shank, jerk-measures are highly informative at every speed, especially over the y-axis, even though N_MSJ and N_MAJ trends are also clear over the x-axis. Finally, whereas vertical velocity and z-axis fluency metrics show clear indications at 6 km/h, KV is more impactful at slower speeds (3-4 km/h). Some of the results for the walking scenarios are illustrated in Figure 13.

Regularity measures have been also obtained for all the scenarios (Figures 14-16). While y-axis regularity and x-axis one were observed in the shank acceleration and angular rate, respectively, during the hamstring curl, only the thigh y-axis regularity over the gyroscope was reliable for the flexion/extension. Regularity calculated in the squat scenario provides indications when obtained from the thigh data, in particular the y-axis acceleration, and the x-, y-axis of the angular rate. Conversely, in the gait test, the regularity obtained from the shank are more informative, in particular the acceleration y-, z-axis (for walking at 3 and 6 km/h respectively) and the gyro z-axis (when walking at 4km/h).

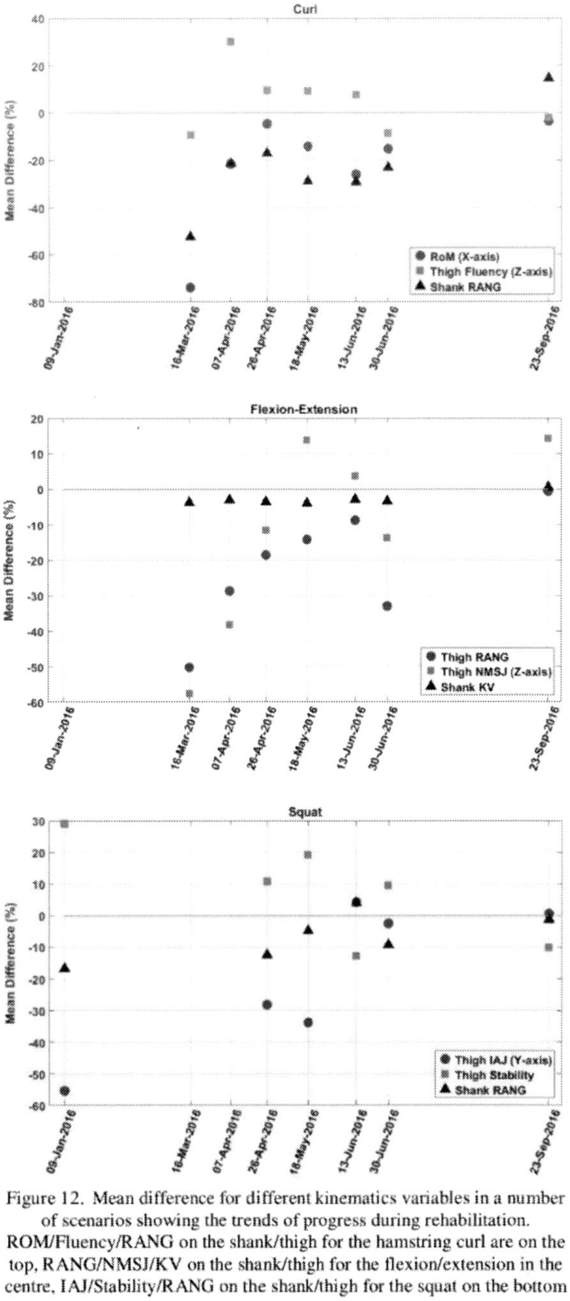

Figure 12. Mean difference for different kinematics variables in a number of scenarios showing the trends of progress during rehabilitation. ROM/Fluency/RANG on the shank/thigh for the hamstring curl are on the top, RANG/NMSJ/KV on the shank/thigh for the flexion/extension in the centre, IAJ/Stability/RANG on the shank/thigh for the squat on the bottom

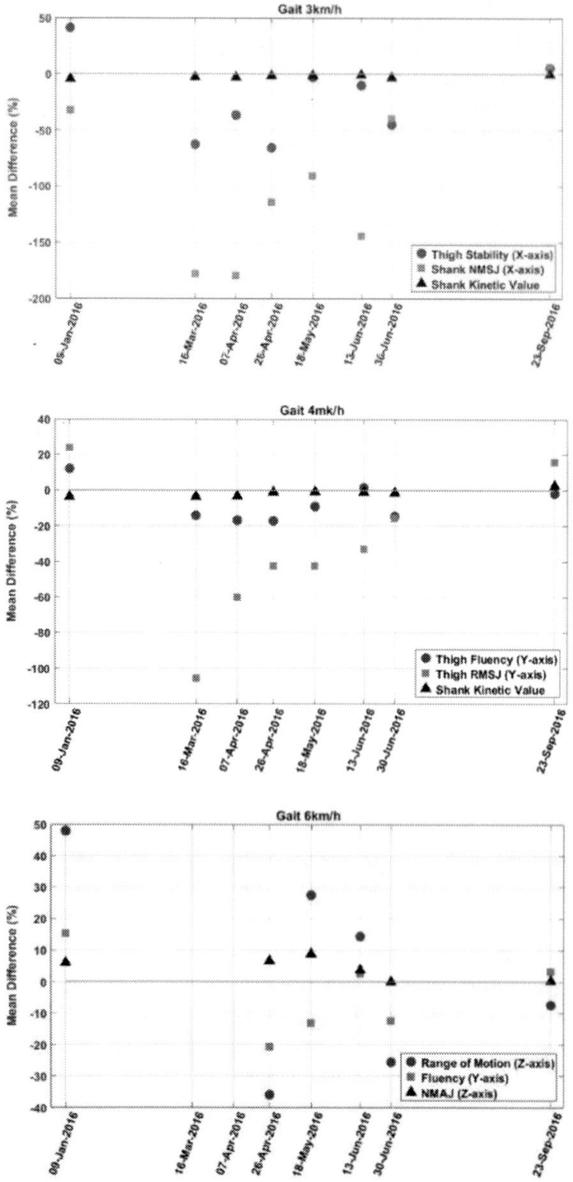

Figure 13. Mean difference for different kinematics variables in a number of scenarios showing the trends of progress during rehabilitation. Stability/NMSJ/KV on the shank/thigh for 3 km/h gait are on the top, Fluency/RMSJ/KV on the shank/thigh for 4 km/h gait in the centre, ROM/Fluency/NMAJ on the shank/thigh for 6 km/h on the bottom

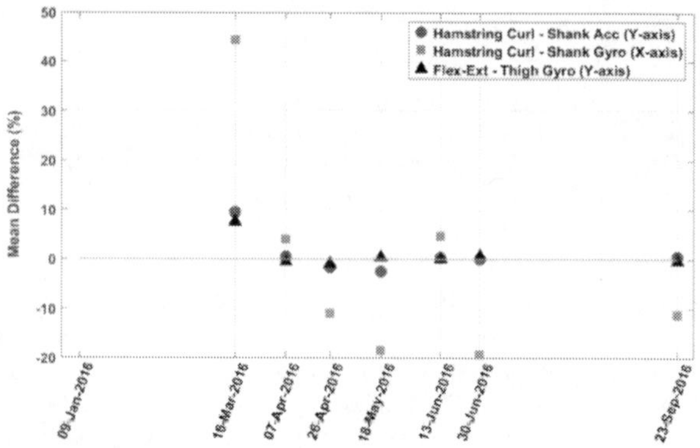

Figure 14. Mean difference for regularity variables in a number of scenarios showing the trends of progress during rehabilitation. Y-axis on the shank acceleration for hamstring curl, x-axis on the gyro shank for hamstring curl, and y-axis on the gyro thigh for flexion/extension are shown. Each variable is normalized according to its mean value for visualization purposes

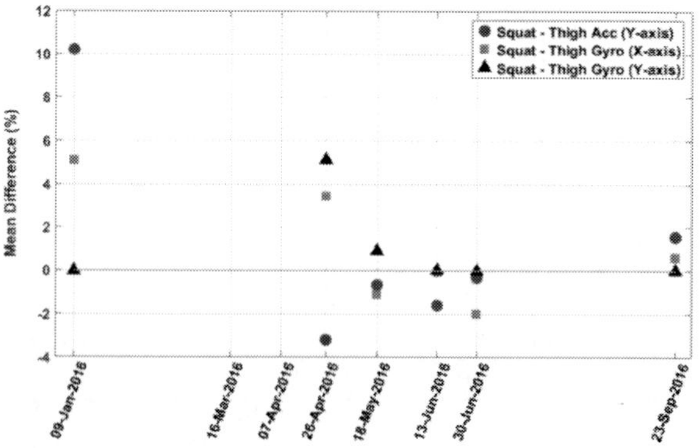

Figure 15. Mean difference for regularity variables for half squat scenario showing the trends of progress during rehabilitation. Y-axis on the thigh acceleration, x-axis on the gyro thigh, and y-axis on the gyro thigh are shown. Each variable is normalized according to its mean value for visualization purposes

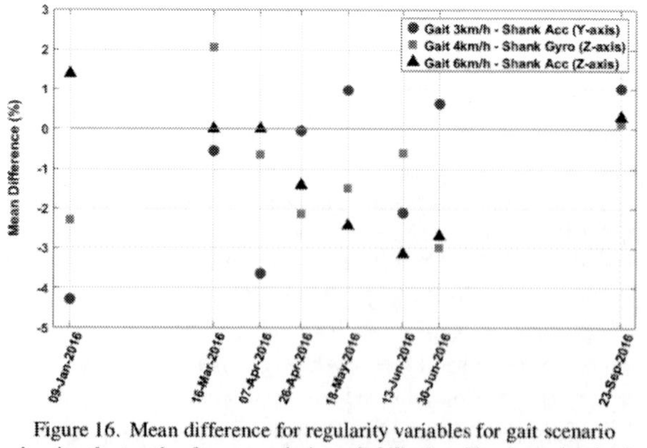

Figure 16. Mean difference for regularity variables for gait scenario showing the trends of progress during rehabilitation. Y-axis on the shank acceleration, z-axis on the gyro shank, and z-axis on the shank acceleration are shown. Each variable is normalized according to its mean value for visualization purposes

To summarize, this work analyzed the body-worn inertial data collected from a patient over the course of rehabilitation defining which features and metrics are the most sensitive for better understanding and monitoring patient's progress in several test. As per gait metrics, temporal variables (and, in particular their CV) can be useful at slower speeds, while clearance-related parameters have higher impact at faster speeds. However, those metrics cannot be adopted for other types of movements which, in turn, can be described through statistical and kinematics variables. Given the movement typically performed on a 2D plane, hamstring curl and flexion/extension show a number of usable metrics obtained from both acceleration and angular rate of the shank and thigh, while the squat is characterized by less features. Probably, this can be due to the fact that the squat test, differently from the other exercises, requires a simultaneous movement of both legs which may involve some compensation between the injured and uninjured limbs not evident when only one leg is involved in the test. Gait can be also illustrated through metrics extrapolated from the inertial data collected on the shank, especially when associated to the sagittal plane. Finally, kinematics variables, in particular ROM, RANG, jerk-based measures, fluency, KV and stability, have proved their sensitivity for a different number of scenarios.

Even though this paper reviewed a large number of features, there remain opportunities for further analysis. This work has not considered frequency-domain, entropy-based, or informatics-theoretic parameters. These parameters should be also evaluated in future studies with the aim of developing a complete framework for collecting data and monitoring patients' progress over the course of rehabilitation. Moreover, as only one subject has been studied for the present proof of concept study, an enhanced number of athletes, with homogeneous characteristics, will also be tested in future so as to have a more robust base for the study and further validate the drawn conclusions in statistical terms. Additional clinical trials are, thus, currently being planned.

VI. CONCLUSIONS

This work presented a wearable inertial system equipped with both hardware and time-domain data analytics for an objective assessment of lower-limbs in patients over the course of rehabilitation. The analysis involved body-worn inertial data collected from thighs and shanks, and the implementation of a number of gait-related, statistical, and kinematics features available in literature, with the ultimate goal of defining which metrics are the most sensitive for better understanding and monitoring patient's progress. Accurate results have been achieved in a number of scenarios.

An enhanced number of subjects, with homogeneous characteristics, will also be tested in future so as to have a more robust base for the study and further validate the drawn conclusions in statistical terms. Additional clinical trials are, thus, currently being planned.

However, the present study proved that inertial-based time-domain features can be used for a quantitative biomechanics monitoring and assessment of lower-limbs in different tests over the course of a nine month rehabilitation program, also defining which of those features should be taken into account by clinicians during their analysis.

REFERENCES

[1] S. Tedesco, A. Urru, J. Peckitt, B. O'Flynn, "Wearable inertial sensors as a tool for quantitative assessment of progress during rehabilitation," Int Conf on Global Health Challenges, pp. 56-59, 2016.

[2] R. Bartlett, "Introduction to sports biomechanics: analysing human movement patterns," (2nd ed.) Routledge, London (2007).

[3] http://www.vicon.com/ (accessed 27/02/2017)

[4] http://www.optitrack.com/ (accessed 27/02/2017)

[5] http://www.codamotion.com/ (accessed 27/02/2017)

[6] E. Mirek, M. Rudzińska, A. Szczudlik, ''The assessment of gait disorders in patients with Parkinson's disease using the three-dimensional motion analysis system Vicon,'' Neurol Neurochir Pol., vol. 41, no. 2, pp. 128-133, 2007.

[7] A.D. Segal, M.S. Orendurff, G.K. Klute, M.L. McDowell, J.A. Pecoraro, J. Schofer, J.M. Czerniecki, ''Kinematic and kinetic comparisons of transfemoral amputee gait using C-Leg and Mauch SNS prosthetic knees,'' J Rehabil Res Dev., vol. 43, no. 7, pp. 857-870, 2006.

[8] S. Lustig, R. A. Magnussen, L. Cheze, P. Neyret, ''The KneeKG system: a review of the literature,'' Knee Surg Sports Traumatol Arthrosc., vol. 20, no. 4, pp. 633-638, 2012.

[9] B. Shabani, D. Bytygi, S. Lustig, L. Cheze, C. Bytygi, P. Neyret, ''Gait knee kinematics after ACL reconstruction: 3D assessment,'' Int Orthopaedics (SICOT), vol. 39, no. 6, pp. 1187-1193, June 2015.

[10] T. Chau, S. Young, S. Redekop, "Managing variability in the summary and comparison of gait data," J Neuroeng Rehabil., vol. 2, no. 22, pp. 1-20, 2005.

[11] A. Castelli, G. Paolini, A. Cereatti, M. Bertoli, U. Della Croce, "Application of a markerless gait analysis methodology in children with cerebral palsy: Preliminary results," Gait & Posture, vol. 42, supp. 2, pp. S4-S5, 2015.

[12] J.F. Item-Glatthorn, N.C. Casartelli, N.A. Maffiuletti, "Reproducibility of gait parameters at different surface inclinations and speeds using an instrumented treadmill system," Gait & Posture, vol. 44, pp. 259–264, 2016.

[13] H.B. Menz, M.D. Latt, A. Tiedemann, M. Mun San Kwan, S.R. Lord, "Reliability of the GAITRite® walkway system for the quantification of temporo-spatial parameters of gait in young and older people," Gait & Posture, vol. 20, no. 1, pp 20–25, 2004.

[14] M. Piirtola, P. Era, "Force platform measurements as predictors of falls among older people – A review," Gerontology, vol. 52, no. 1, pp. 1-16, 2006.

[15] N.J. Collins, D. Misra, D.T. Felson, K.M. Crossley, E.M. Roos, "Measures of knee function," Arthritis Care Res (Hoboken), vol. 63, no. 11, pp. S208–S228, 2011.

[16] S. Tilley, N. Thomas, "What knee scoring system?," J. Bone Joint Surgery, pp. 1-4, 2010.

[17] D.T.P. Fong, Y.Y. Chan, "The use of wearable inertial motion sensors in human lower limb biomechanics studies: A systematic review," Sensors, vol. 10, no. 12, pp. 11556-11565, 2010.

[18] J. Barton, A. Gonzalez, J. Buckley, B. O'Flynn, S.C. O'Mathuna, "Design, fabrication and testing of miniaturised wireless inertial measurement units (IMU)," IEEE Electronic Components and Technology Conference (ECTC), pp. 1143-1148, 2007.

[19] M. Walsh, M. O'Grady, M. Dragone, R. Tynan, A. Ruzzelli, J. Barton, B. O'Flynn, G. O'Hare, C. O'Mathuna, "The CLARITY modular ambient health and wellness measurement platform," IEEE Int. Conf. Sensor Technologies and Applications (SENSORCOMM), pp. 577-583, 2010.

[20] D. Kennedy, M. Walsh, B. O'Flynn, "Development of a miniature, low-cost wave measurement solution," IEEE Oceans-St. John's, 2014, pp. 1-6.

[21] B. O'Flynn, J.T. Sanchez, S. Tedesco, B. Downes, J. Connolly, J. Condell, K. Curran, "Novel smart glove technology as a biomechanical monitoring tool," Sensors & Transducers, vol. 193, no. 10, pp. 23-32, Oct. 2015.

[22] N. Ahmad, R. Ariffin, N.M. Khairi, V. Kasi, "Reviews on various inertial measurement unit (IMU) sensor applications," Int J Signal Processing Systems, vol. 1, no. 2, pp. 256-262, 2013.

[23] K. Aminian, B. Najafi, C. Büla, P.F. Leyvraz, P. Robert, "Spatiotemporal parameters of gait measured by an ambulatory system using miniature gyroscopes," J Biomech, vol. 35, pp. 689–699, 2002.

[24] R.E. Mayagoitia, A.V. Nene, P.H. Veltink, "Accelerometer and rate gyroscope measurement of kinematics: an inexpensive alternate to optical motion analysis systems," J Biomech, vol. 35, pp. 537–542, 2002.

[25] A.M. Sabatini, C. Martelloni, S. Scapellato, F. Cavallo, "Assessment of walking features from foot inertial sensing," IEEE Trans. Biomedical Eng., vol. 52, no. 3, pp. 486-494, 2005.

[26] E. Wade, M.J. Mataric, "Design and testing of lightweight inexpensive motion-capture devices with application to clinical gait analysis," Int. Conf. Pervasive Computing Technologies Healthcare, pp. 1-7, April 2009.

[27] L. Atallah, A. Wiik, G. Jones, B. Lo, J. Cobb, A. Amis, G.Z. Yang, "Validation of an ear-worn sensor for gait monitoring using a force-plate instrumented treadmill," Gait & Posture, vol. 35, pp. 674–676, 2012.

[28] S. Kobashi, K. Kawano, Y. Tsumori, S. Yoshiya, Y. Hata, "Wearable joint kinematic monitoring system using inertial and magnetic sensors," IEEE Workshop Robotic Intelligence Informationally Structured Space, pp. 25–29, 2009.

[29] J.C. Van den Noort, V.A. Scholtes, J. Harlaar, "Evaluation of clinical spasticity assessment in cerebral palsy using inertial sensors," Gait & Posture, vol. 30, no. 2, pp. 138–143, 2009.

[30] Y. Guo, D. Wu, G. Liu, G. Zhao, B. Huang, L. Wang, "A low-cost body inertial-sensing network for practical gait discrimination of hemiplegia patients," Telemed J E Health, vol. 18, no. 10, pp. 748-754, 2012.

[31] Y. Guo, G. Zhao, G. Liu, Z. Mei, K. Ivanov, L. Wang, "Balance and knee extensibility evaluation of hemiplegic gait using an inertial body sensor network," BioMedical Engineering OnLine, vol. 12, pp. 83, 2013.

[32] S. Das, L. Trutoiu, A. Murai, D. Alcindor, M. Oh, F. De La Torre, J. Hodgins, "Quantitative measurement of motor symptoms in Parkinson's disease: A study with full-body motion capture data," IEEE Int Conf Eng Med Biol Soc, pp. 6789–6792, 2011.

[33] J.R. Merory, J.E. Wittwer, C.C. Rowe, K.E. Webster, "Quantitative gait analysis in patients with dementia with Lewy bodies and Alzheimer's disease," Gait & Posture, vol. 26, no. 3, pp. 414 –419, 2007.

[34] M. Monda, A. Goldberg, P. Smitham, M. Thornton, J. McCarthy, "Use of inertial measurement units to assess age-related changes in gait kinematics in an active population," J Aging Phys Act, vol. 23, no. 1, pp. 18-23, 2015.

[35] S. Tedesco, A. Urru, A. Clifford, B. O'Flynn, "Experimental validation of the Tyndall portable lower-limb analysis system with wearable inertial sensors," Procedia Engineering, The Engineering of Sport 11, vol. 147, pp. 208-213, 2016.

[36] R. Nerino, L. Contin, A. Tirri, G. Massazza, A. Chimienti, G. Pettiti, N. Cau, V. Cimolin, "An improved solution for knee rehabilitation at home," Int Conf Body Area Networks, pp. 62-68, 2014.

[37] P. Fergus, K. Kafiyat, M. Merabti, A. Taleb-bendiab, A. El Rhalibi, "Remote physiotherapy treatments using wireless body sensor networks," Int Conf Wireless Communications Mobile Computing: Connecting the World Wirelessly, pp. 1191–1197, 2009.

[38] S.L. Han, M.J. Xie, C.C. Chien, Y.C Cheng, C.W. Tsao, "Using MEMS-based inertial sensor with ankle foot orthosis for telerehabilitation and its clinical evaluation in brain injuries and total knee replacement patients," Microsystem Technologies, vol. 22, no. 3, pp. 625-634, 2016.

[39] O.M. Giggins, K.T. Sweeney, B. Caulfield, "Rehabilitation exercise assessment using inertial sensors: a cross-sectional analytical study," J NeuroEng Rehab, vol. 11, pp. 158, 2014.

[40] P.E. Taylor, G.J. Almeida, T. Kanade, J.K. Hodgins, "Classifying human motion quality for knee osteoarthritis using accelerometers," IEEE Int Conf Eng Med Biol Soc, pp. 339-343, 2010.

[41] P.E. Taylor, G.J. Almeida, J.K. Hodgins, T. Kanade, "Multi-label classification for the analysis of human motion quality," IEEE Int Conf Eng Med Biol Soc, pp. 2214-2218, 2012.

[42] E. Papi, D. Osei-Kuffour, Y.M.A. Chen, A.H. McGregor, "Use of wearable technology for performance assessment: A validation study," Med Eng Phys, vol. 37, no. 7, pp. 698-704, 2015.

[43] G. Fenu, G. Steri, "IMU based post-traumatic rehabilitation assessment," Int. Symp. Applied Sciences in Biomedical and Communication Technologies , pp. 1-5, 2010.

[44] P.C. Chen, C.N. Huang, I.C. Chen, C.T. Chan, "A rehabilitation exercise assessment system based on wearable sensors for knee osteoarthritis," Int Conf Smart Homes Health Telematics, pp. 267-272, 2013.

[45] K.H. Chen, P.C. Chen, K.C. Liu, C.T. Chan, "Wearable sensor-based rehabilitation exercise assessment for knee osteoarthritis," Sensors, vol. 15, no. 2, pp. 4193-4211, 2015.

[46] Y.C. Tseng, C.H. Wu, F.J. Wu, C.F. Huang, C.T. King, C.Y. Lin, J.P. Sheu, C.Y. Chen, C.Y. Lo, C.W. Yang, C.W. Deng, "A wireless human motion capturing system for home rehabilitation," Int Conf Mobile Data Management: Systems, Services and Middleware, pp. 359–360, 2009.

[47] V. Bonnet, C. Mazzà , P. Fraisse, A. Cappozzo, "A least-squares identification algorithm for estimating squat exercise mechanics using a single inertial measurement unit," J. Biomech., vol. 45, no. 8, pp. 1472–1477, 2012.

[48] V. Bonnet, C. Mazzà , P. Fraisse, A. Cappozzo, "Real-time estimate of body kinematics during a planar squat task using a single inertial measurement unit," IEEE Trans. Biomed. Eng., vol. 60, no. 7, pp. 1920–1926, 2013.

[49] V. Bonnet, V. Joukov, D. Kulic, P. Fraisse, N. Ramdani, G. Venture, "Monitoring of hip and knee joint angles using a single inertial measurement unit during lower limb rehabilitation," IEEE Sensors Journal, vol. 16, no. 6, pp.1557-1564, 2016.

[50] O. Giggins, D. Kelly, B. Caulfield, "Evaluating rehabilitation exercise performance using a single inertial measurement unit," Int Conf Pervasive Computing Technologies for Healthcare, 2013.

[51] J. Brutovsky, D. Novak, "Low-cost motivated rehabilitation system for post-operation exercises," IEEE Int Conf Eng Med Biol Soc, pp. 6663–6666, 2006.

[52] D. Fitzgerald, J. Foody, D. Kelly, T. Ward, C. Markham, J. McDonald, B. Caulfield, "Development of a wearable motion capture suit and virtual reality biofeedback system for the instruction and analysis of sports rehabilitation exercises," Int Conf IEEE Eng Med Biol Soc, pp. 4870-4874, Aug. 2007.

[53] S. Ananthanarayan, M. Sheh, A. Chien, H. Profita, K. Siek, "PT Viz: towards a wearable device for visualizing knee rehabilitation exercises," SIGCHI Conf Human factors Computing Systems, pp. 1247-1250, 2013.

[54] G. Lullini, L. Berti, M. Ortolani, A. Leardini, A. Valsecchi, "Are modern inertial sensor and bio-feedback based rehabilitation systems reliable enough? A validation study by standard gait analysis," Gait & Posture, vol. 39, supp. 1, pp. S42-S43, 2014.

[55] A. Leardini, G. Lullini, S. Giannini, L. Berti, M. Ortolani, P. Caravaggi, "Validation of the angular measurements of a new inertial-measurement-unit based rehabilitation system: comparison with state-of-the-art gait analysis," J NeuroEngineering Rehab, vol. 11, pp. 136, 2014.

[56] J.F.S. Lin, D. Kulic, "Human pose recovery for rehabilitation using ambulatory sensors," IEEE Int Conf Eng Med Biol Soc, pp. 4799-4802, 2013.

[57] M. Field, D. Stirling, M. Ros, Z. Pan, F. Naghdy, "Inertial sensing for human motor control symmetry in injury rehabilitation," IEEE/ASME Int Conf Advanced Intelligent Mechatronics, pp. 1470-1475, 2013.

[58] R. Houmanfar, M. Karg, D. Kulic, "Movement analysis of rehabilitation exercises: distance metrics for measuring patient progress," IEEE Systems Journal, vol. 10, no. 3, pp. 1014-1025, 2016.

[59] https://www.tyndall.ie/wireless-sensor-networks-2 (accessed 27/02/2017)

[60] R.M. Nilssen, J.L. Helbostad, "Estimation of gait cycle characteristics by trunk accelerometry," J Biomechanics, vol. 37, no. 1, pp. 121-126, 2004.

[61] M. Zhang, B. Lange, C.Y. Chang, A.A. Sawchuk, A.A. Rizzo, "Beyond the standard clinical rating scales: fine-grained assessment of post-stroke motor functionality using wearable inertial sensors," IEEE Eng Med Biol Soc, pp. 6111-6115, 2012.

[62] C. Strohrmann, R. Labruyere, C.N. Gerber, H.J. van Hedel, B. Arnrich, G. Troster, "Monitoring motor capacity changes of children during rehabilitation using body-worn sensors," J NeuroEngineering Rehab, vol. 10, pp. 83, 2013.

[63] N. Hogan, D. Sternad, "Sensitivity of smoothness measures to movement duration, amplitude, and arrests," J Mot Behav, vol. 41, no. 6, pp. 529-534, 2009.

[64] L. Schwickert, R. Boos, J. Klenk, A. Bourke, C. Becker, W. Zijlstra, "Inertial sensor based analysis of lie-to-stand transfers in younger and older adults," Sensors, vol. 16, no. 8, pp. 1277, 2016.

[65] S. Jiang, B. Zhang, D. Wei, "The elderly fall risk assessment and prediction based on gait analysis," IEEE International Conference on Computer and Information Technology, pp. 176-180, 2011.

[66] M. Gietzelt, G. Nemitz, K.H. Wolf, H.M.Z. Schwabedissen, R. Haux, M. Marschollek, "A clinical study to assess fall risk using single waist accelerometer," Journal Informatics for Health and Social Care, vol. 34, no. 4, 2009.

[67] https://uk.mathworks.com/products/matlab.html (accessed 27/02/2017)

[68] T. Seel, J. Raisch, T. Schauer, "IMU-based joint angle measurement for gait analysis," Sensors, vol. 14, no. 4, pp. 6891-6909, 2014.

Development and Aging on Motor Control Function with Precise Observations of Synchronization Hands' Movements

Kyota Aoki and Hisanori Hotta
Graduate school of Engineering, Utsunomiya University, Utsunomiya, Japan
e-mail: kyota@is.utsunomiya-u.ac.jp hisanori6432@yahoo.co.jp

Abstract—**In advanced countries, populations are getting older. Cognitive disorder is a large problem in the countries. To understand the cognitive disorder with aging, we need to have a whole image of the development and the aging of our cognitive functions. We need to measure the performance of brain functions with the process of development and aging. For synchronizing with other motion, we need to feel the other motion, to recognize, to memorize, and to generate the synchronizing motion. We need many kinds of brain functions to make synchronizing movement. The authors proposed the cooperative visual synchronization task, its measuring method, implementation and experiments to measure and evaluate the performance of motor control function. The new task and the measuring method enable to measure the precise movements and evaluate the performance of motor control function easily in a short period. The proposed method is safe, because there is no need to attach the device to a subject nor to make exaggerated motions. This paper presents the results of experiments about primary school pupil, high-school students, young students, manhood people and elderly people. In addition, the authors show the overview of development and aging process of motor control function from objective measurement of cooperative movement in both hands.**

Keywords-aging; development; aging process; motor control function; measurement; evaluation.

I. INTRODUCTION

Our brain function starts its development from the age of infant, keeps its performance in manhood, and deteriorates with aging. There is no overview about our brain performance from infant to elderly people in single measure. IQ test can be applied to wide range of ages. However, the IQ test needs much time. The IQ test is not the same in all range of ages. The IQ test for an infant is not same with the one for a young adult.

With aging, our physical function deteriorates, and also our brain functions [1]. In advanced countries, the populations are getting older. Our physical deteriorations are measured easily. Also, we need to measure the deterioration of brain functions. There are tests to measure the memory functions and the cognition disorders.

A cooperative movement with other movements is more difficult than simple movements. For instance, clapping hands is easy. However, clapping hands synchronizing with other people is difficult. Synchronizing movement is the base of cooperative movement. For synchronizing with another motion, we need to feel the other motion, to recognize, to memorize, and to generate the synchronizing motion. We may estimate the performance of total brain function observing the process of synchronizing movement. Because of it, we need to work many brain functions to complete the synchronizing movement.

Many motor tasks measure the performance of human motor function. They are the Purdue pegboard task, a seal affixation task, a tray-carrying task, etc. [2] - [4]. These tasks estimate the performance of human motor function based on the results from the tasks. There is no observation on the process of completing the tasks. Some synchronization tasks measure motor function of a human. One example is a synchronization of finger taps with periodically flashing visual stimuli and synchronization with an auditory metronome. In these tasks, the timing between the stimuli and the tapping is measured. There is no observation about the process of the tapping [5] - [11].

Recently, many cheap and easy measurement methods for the movements of a human body have been developed. For instance, some of these sensors include Kinect sensor, and Leap motion sensor [12] [13]. Many applications use these sensors to control computers. For example, many video games use these sensors to control avatars in the games [14].

Using the new motion sensor, we can measure the motion of hands easily and precisely. The human hands are the parts of a body that can make the most complex movements. We have proposed a method that measures the precise movements of hands synchronizing the movements of hands on a display. The synchronization needs visual perception of the displayed hands' images and precise control over the arm muscles. The resulting measure is very sensitive. With this measure, we can observe the performance of the motor function precisely [15].

This paper proposes the overview of development and aging process of our brain with a new estimation method to evaluate the performance of a brain function by the measurement of a motion control function in cooperative synchronizing movements. To draw the outline of the development and the aging process, this paper needs many age groups that spans from infant to elderly people. This paper includes the experiments about primary school pupil, high school students, young, manhood, and elderly people. In this paper, there is no infant and junior high school student. However, this paper can show the outline of development and aging of brain function in a single measure.

The rest of this paper is organized as follows. Section II discusses a task to synchronize hands' movements with visual presentation. Next, we discuss the experimental setup in

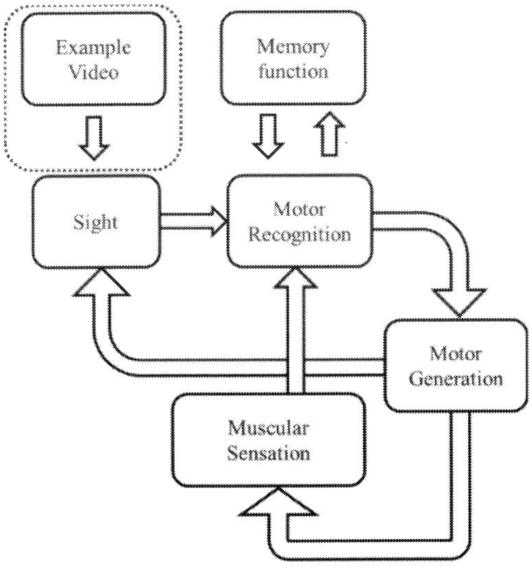

Figure 1. Relations among functions.

Section III, and show our experimental results in Section IV. Finally, we conclude this work in Section V.

II. Visual Synchronization Task

A. Task and Brain Function

Many motor tasks intend to measure the performance of the motor function of a human [2] - [4]. Motor function is not a strength of muscles. The strength of muscles is one measure of our body performance. We may have problems of our muscles in the case that we cannot have a good performance of motor functions. For the people with healthy body, their brain performance decides the performance on motor tasks. Those motor tasks measure the performance of some motions. However, most of these tasks measure the results from the tasks. They do not measure the process of motions directly.

Some tasks measure the synchronization between a finger tap and stimuli [5] - [11]. With human observations, it is difficult to measure the process of synchronizing movements. Classical works measure the timing of pushing a switch. Now, we can use a Kinect sensor and a Leap Motion sensor. These sensors measure the three-dimensional movements of a human body. With these sensors, we can measure the precise movements of a human body.

We can synchronize our movements with each other. For instance, when dancing in groups, dancers can synchronize their movements with each other. A synchronization of movement is a more difficult work than a simple imitation of movement. To generate synchronizing movements, we need to observe the motion synchronized. We need to generate the motion to be similar to the motion synchronized. We need to observe the generated motion synchronizing the original motion. We need to estimate the divergence between the original motion synchronized and the motion synchronizing the original motion. We need to control the speed of the motion synchronizing. These functions form a feedback loop. However, there is a delay in our processing. To compensate our brain's processing delay, we need to estimate the delay itself and make a proper amount of feedforward. Therefore, our brain functions have a feedback loop and a feedforward loop.

This processing loop is shown in Fig. 1. For estimating the total brain function, we need to include all the functions of the brain. In this paper, we call the task that a subject synchronizes one's movement to the displayed movement, as the visual synchronization task. The visual synchronization task includes vision and motor functions. The vision includes not only the static sight, but also the dynamic sight.

The visual synchronization task is more difficult than audio synchronization. Therefore, we observe the wider brain functions with the visual synchronization tasks than the audio synchronization tasks.

Our proposed visual synchronization task is the synchronization between the pose of stimuli on a display and the pose of the hands. Our synchronization task is not the synchronization between the timing of the stimuli and the timing of the action. The measurement of timing gets only one scalar value in a cycle of stimuli. In our proposed synchronization task, the measuring result is a sequence of the triples of the pose of subject's hands in a duration of stimuli. For instance, we have 100 triples of floating numbers in a second.

B. Stimuli of Visual Synchronization Task

For the motor task, the authors select the rotation of both hands. Rotation is a difficult movement with a hand. For analyzing the synchronization easily, the authors make the stimuli that rotation angles follow a precise sine curve. Fig. 2 shows the example of a pure sine curve. If stimuli form precise sine curves, we easily evaluate the observed motion comparing with the pure sine curve.

Our stimuli are a displayed video of both hands' rotation. Fig. 3 shows some frames of a hands' rotation video for the visual synchronization task. In a real world, it is difficult to control precisely the motion of hands to follow the pure sine curve. If the stimuli images do not follow the pure sine curve,

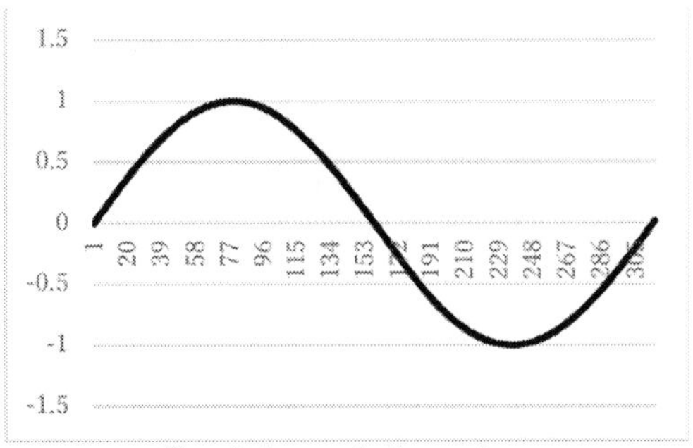

Figure 2. An example of a pure sine curve.

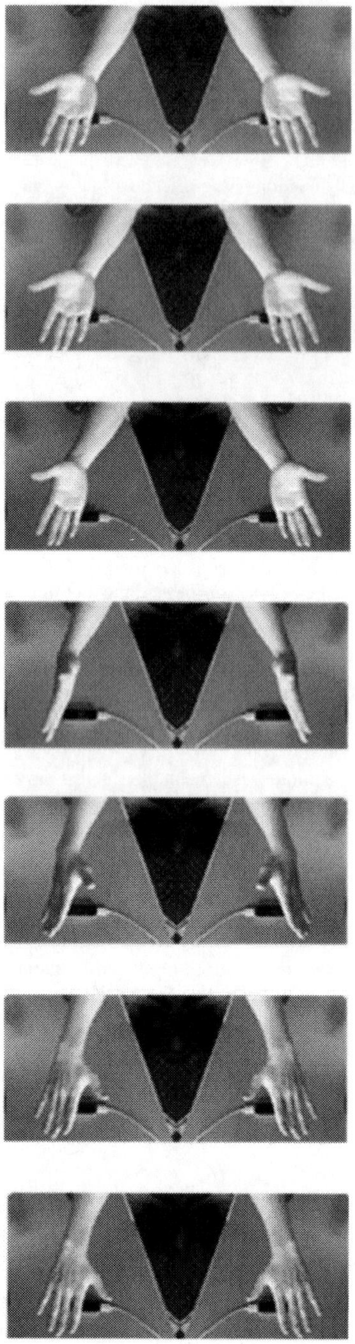

Figure 3. Stimuli Images.

it is also difficult to evaluate the synchronization between the stimuli images and the motions of hands. We can use many measures. For instance, the square error of the poses between stimuli and hands can work as the performance measure of the observed hands' movements. However, the square error changes in the scale of movements of hands. This paper needs the method to measure the performance of the motor control function. The flexibility of our arms must not affect the performance measure. Therefore, square error is not fit for our purpose. Using a pure sine curve as stimuli, we can estimate the performance of our hands' movements with invoking a signal-to-noise ratio based on the communication theory. This proposes a strict base to the proposed measuring method for the performance of the control function.

C. *Implementation*

Fig. 3 shows the sequence of stimuli examples. The total sequence of stimuli includes 67 images. The images are proposed on a display with a constant interval from top to bottom. Then, they are proposed from bottom to top. These two sequences make one cycle of the stimuli of hands' motion. In the stimuli images, the right hand and the left hand are the same pose. The right one is the mirror image of the left one.

The authors propose the stimuli generation method that displays a proper image at the precise timing. The authors recorded the motion of hand's rotation with a camera and the Leap-motion sensor simultaneously using a PC. The records include the precise time stumps. From the recorded images and measured rotation angles, we constructed the sequence of images that rotations follow a pure sine curve.

In displaying images on a display, there is some delay or progress of a timing. Our implementation controls the timing of displaying each frame with selecting the best-fit frame at the timing. As a result, our implementation displays smooth transition between successive frames.

Our implementation has two sub-processes. The one sub-process displays the stimuli images. The other sub-process measures the pose of hands. With this multi-processing, our implementation enables to show the stimuli and to measure the reaction of a subject in the best performance.

In the following experiments, one cycle of hands' rotation is completed in a second. We can rotate our hands 1.5 cycle at a second. However, 1.5 cycle at a second is too fast for many subjects in our previous experiments.

We can rotate our hands much more slowly. However, in slow rotations, we can easily follow the position of displayed hands' image. In the case, we do not synchronize our movements to the displayed hands' movements. We only imitate the poses of the displayed hands. There is no synchronization of movements. Therefore, we cannot use much slower movements.

We measure the pose of hands with the leap-motion sensor [13]. The sensor measures the poses of hands 100 times for a second. As a result, we have two measurements of the poses of hands in each cycle of hands' rotation. The leap-motion sensor measures the three-dimensional pose of a hand. We use only the rotations around forearms.

D. Motion synchronization measure

We define the synchronization measure using Fast Fourier Transform (FFT) results of the measured rotations of poses of both hands in each cycle [16]. If a subject makes complete synchronization to the stimuli, the resulting poses of both hands follow a pure sine curve. As a result, at every cycle of the rotation of hands, the result of FFT has a zero value at the second term or higher terms. We define the measure representing the noise-to-signal ratio as (1). This measure increases with increasing the difference from ideal sine curve.

$$NSM = (\sum_{x=2}^{t/4} m_x) \Big/ m_1 \qquad (1)$$

In (1), t is the number of terms. m_x is the absolute value of the x-th term of the result of FFT. m_1 is the power of the lowest frequency. This represents one cycle of a hand's rotation. If the rotation of a hand follows the stimuli images precisely, m_1 carries all powers of the hand's rotation. Other terms carry no power. In that case, the measure in (1) is zero. The result FFT has much higher terms from $t/4+1$ to t. The t/2 or higher terms are mirrored of lower terms. Therefore, we need only treat a $t/2$ or lower terms. Under $t/2$ terms, there are many noises in upper terms. In a normal processing of FFT, we use window function to decrease the noises in observations. However, the authors define the measure at each cycle. Therefore, it is difficult to use window function. Therefore, the authors use only lower half of the all terms. If we can have 100 measurements in a cycle, we use 25 terms. They represent the motion at every 1/50 seconds. This is enough precise for observing our brain control loop.

m_0 is a value that represents the average of poses. This is not included in (1). As a result, this measure does not depend on the absolute poses of hands.

Some people rotate their hands largely. Others do not. (1) is the ratio between a signal and a noise. Therefore, the scale of rotations does not affect the measure. We call this measure as Non-Smoothness-Measure (NSM). This measure may span from zero to infinity.

Our proposed system observes two hands. Therefore, at every cycle, we have a pair of NSMs.

E. Phase

The NSM is the measure of the difference of a motion from the displayed motion. However, there is a difference of timing between the displayed motion and a user's motion. The tapping test measures the difference between a stimulus and the response of a user. In the proposed synchronization task, the difference in timing is the difference of phases.

In the result of FFT, there are phases of all frequencies. 0-th term carries a constant pose. Therefore, it has not a phase. The first term represents the signal of the stimuli. Therefore, the authors use the phase of the first-term to estimate the phase of the motion of a hand.

In our experiments, these are from 1 Hz to 50 Hz of terms of FFT. The signal of 1 Hz represents the ideal motion based on the proposed example motion. Therefore, we use the phase of the signal of 1 Hz for evaluating the timing of the motion.

III. Experiments and Discussions

A. Experiments Setup

1) Precise type

From the pre-experiments, the speed of the hands' rotation is best at one cycle per second. Subjects need about three cycles to synchronize their movements of hands to the proposed motion images and remember the motion. For measuring a stable result, we need at least three cycles. As a result, one trial of an experiment needs six cycles at least. However, there may be some error measurements. For stable measurements, we need some redundancy. The authors decide 10 cycles to measure the hands' movements synchronizing the displayed stimuli images. After the stimuli images disappear, the authors want to observe the decay of the motion memory. Therefore, the authors observe the motion of hands in fifteen seconds. There is no ground about the length of observation. However, a subject feels 30 seconds of a trial to be very long. For getting reliable results, we decide that the length of a trial is 25 cycles of rotations. This means that one trial needs 25 S. Fig. 4 shows the relations among parts, cycles, and sections in a trial. A cycle is one flip of hands. There are two parts in a trial. One part is an example displaying part. The other is no-example displaying part. The sections are periods to analyze measured data. The first section shows the status of a subject in the motion example displaying part. The second section does the status just after the disappearance of the motion example. The third section shows the status at some seconds after the example motion disappears. Before starting a trial, we instruct subjects to synchronize their hands to the displayed hands' images and continue to move the hands after the example motion disappears.

2) Simple type

In a large-scale measurement, the precise type measurements is too long. It needs at least 25 sends to complete. Moreover, there are many error measurements. Because, many subjects cannot keep their hands' movements

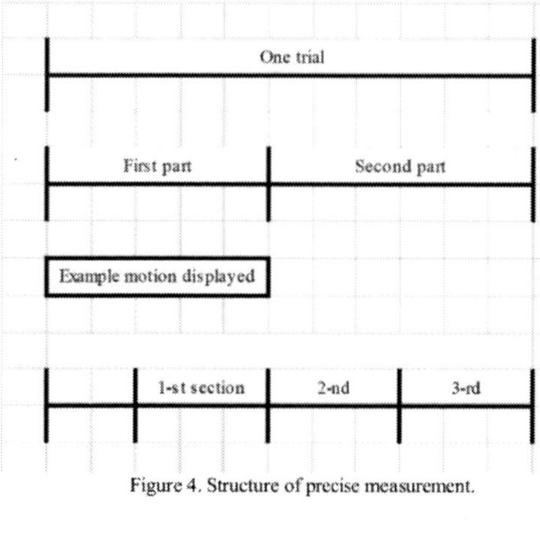

Figure 4. Structure of precise measurement.

after the displayed hands' movements disappear. In a large-scale measurement, the total length of measurements must be long. In a long term, there must be some changes of environment. In the case, the ratio of error measurement increases. However, the authors hope to have a valid trial with some error measurements. Therefore, the authors define the trial that includes only the example displayed part. The authors increase the length of the example displaying part increase to fifteen seconds. This is a simple type trial.

Our simple type of visual synchronization task has no measurements of poses of hands without displayed hands' movements. To measure enough valid data, the length of measurement is 15 cycles of hands' rotations. A trial needs only 15 seconds. It is 10 seconds shorter than the precise type measurement. With longer displayed hands' movement images, many subjects succeed to keep their hands synchronizing to the displayed hands' movements. With some error measurements' cycles, we can have enough valid measurements. Therefore, we have a large success ratio of trials.

B. Experiment

1) People in primary school ages

We obtained about 400 trials with whole pupils in a normal public primary school in Japan. Their ages span from 7 years old to 13 years old. The pupils with heavy healthy problems attend special support schools. Therefore, the pupils in a normal primary school are healthy. The pupils in a first grade in a primary school have difficulties to follow the instructions of the precise type measurement. The precise type measurement needs 1.7 times as much time as the simple type measurement. We cannot withstand this increase of measurement time. The measurement type must be the simple one.

2) Hight school Students

There are 35 subjects. However, there are many error trials. We have 24 valid measurements. They are third grade of high school. They are 17 or 18 years old. They are 21 male students and 3 female students. The authors made VST trials for the high-school students in voluntary basis. Therefore, the authors selected the simple type measurements.

3) Young people (University Students)

We obtained 232 trials with five healthy male students, with ages between 22 - 24 years old. The measurement type is the precise one.

4) Manhood people

We performed experiments with manhood people. They are 3 females and 12 males. They are from 25 years old to 63 years old. Their average age is 44 years old. They are all healthy. They made 168 trials. The measurement type is the simple one.

5) Elderly people

We performed experiments with elderly people, 75 years old in average. They are from 66 years old to be 82 years old. They are all healthy in their ages. In our observation, one female has a difficulty about walking. Therefore, we have 14 healthy elderly people, 4 males and 10 females.

The measurement type is precise one. Each person made two trials. Therefore, we have the 28 trials. One trial had a failure in measurement. We obtained 27 valid measurements of the trials.

6) Performance measure for a trial

a) Precise type measurement

At each trial, we have 25 pairs of NSMs and 25 pairs of phases at most. In many cases, a subject could not move his hands as the displayed hands at the first cycle. The NSM shows the difference of the motion of subject's hands from the proposed example motion at each cycle. The phase represents the difference of the timing between the proposed example motion and the motion made by a subject.

In a single cycle, measured movements of hands may match the proposed example movements accidentally. We estimate the performance of the motion control function with the average motions in three continuous cycles. In addition, we estimate the performance of a subject in a trial with the best movements in the averages of three continuous cycles.

Equation (2) defines the performance of a hand in a trial.

$$\mathrm{NSMHp} = \min_{i=1,8} average(NSM_i, NSM_{i+1}, NSM_{i+2}) \quad (2)$$

NSMHp is the performance of a hand in a trial in precise type. NSM_i is the *NSM* at i-th cycle defined as (1). We have two NSMHps in a trial. They represent the performances of both hands.

We define the performance measure in a trial as (3).

$$\mathrm{NSMTp} = \min(NSMHp_L, NSMHp_R) \quad (3)$$

In (3), NSMTp is the performance measure in a trial. $NSMHp_L$ is the NSMHp of the left hand. $NSMHp_R$ is the NSMHp of the right hand. This NSMTp represents the performance of a subject in a trial. Our previous experiments show that the difference between both hands is small. However, we select a better one for the measure of motor control function. A subject shows a difference of the performance between a left hand and a right hand. This paper uses the better performance.

b) Simple type measurements

At each simple type measurement, there are 15 cycles of hands' movements. At this type of measurement, the author uses all 15 cycles to estimate the performance.

Equation (4) defines the performance of a hand in a trial.

$$\mathrm{NSMHs} = \min_{i=1,13} average(NSM_i, NSM_{i+1}, NSM_{i+2}) \quad (4)$$

NSMHs is the performance of a hand in a trial. NSM_i is the *NSM* at i-th cycle defined as (1). We have two NSMHs in a trial. They represent the performances of both hands.

$$\mathrm{NSMTs} = \min(NSMHs_L, NSMHs_R) \qquad (5)$$

In (5), NSMTs is the performance measure in a trial. $NSMHs_L$ is the NSMHs of the left hand. $NSMHs_R$ is the NSMHs of the right hand. The NSMTs represents the performance of a subject in a trial. However, we select a better one for the measure of motor control function. A subject shows a difference of the performance between a left hand and a right hand. This paper uses the better performance.

There is a difference between the definition of NSMTp and the definition of NSMTs. However, in most trials, around fifth cycle, both of NSMHs and NSMHp show the least value. Therefore, we can treat NSMTp and NSMTs in a same manner. In the precise type measurements, we have multiple measurements of a subject. The precise measurements are well controlled.

IV. Result and Discussions

A. NSMs

1) Pupil in Primary school

The rotation of hands can be achieved from the five years old. In Japan, pupils in primary school are from six years old to twelve years old. In these ages, the performance of a brain represented as motor control function are developing. The authors have a chance to measure the NSMs of all pupils in a primary school.

We summarize the measurements with subjects' age. Fig. 5 shows the distribution of all measurements in a primary school. In Fig. 5, x-axis is the months from birth. The y-axis is NSMTs. A linear approximation of the distribution is shown (6).

Table I shows the averages and the standard derivations in each school year. With the progress of school year, the average NSMTs decreases. We can estimate linear approximation of the distribution. (6) shows the linear approximation. In a month, a pupil in a primary school shows 0.0021 decrease of NSMTs.

$$\mathrm{NSMT} = -0.0021\mathrm{M} + 0.539 \qquad (6)$$

In (6), M is the month from birth of a subject.

The difference of 0.0021 is difficult to be measured. However, VST can distinguish the change of 0.002 NSMT. VST can measure the development in three months.

Table I. Mean and STD of NSMTs at each school year in a primary school.

School year	Average	STD
1	0.501	0.353
2	0.381	0.122
3	0.348	0.117
4	0.301	0.064
5	0.297	0.065
6	0.273	0.063

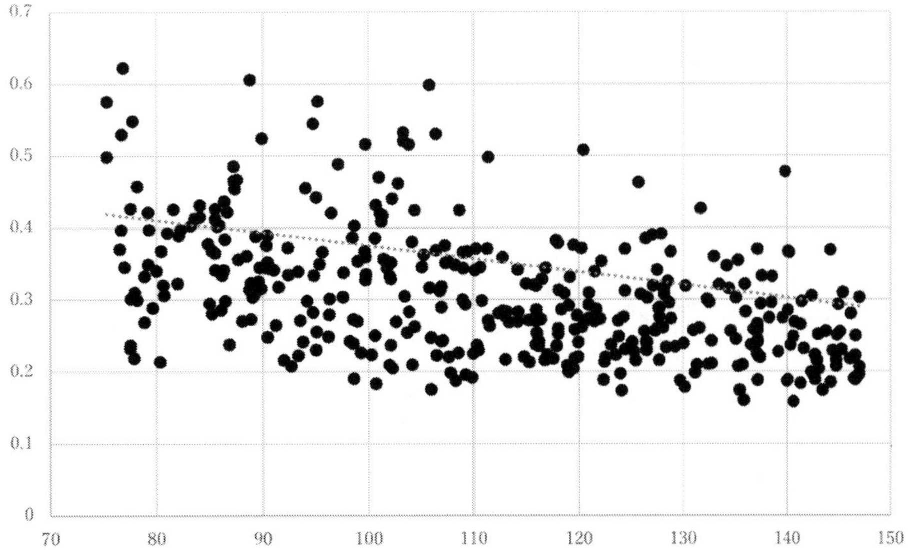

Figure 5. Age-NSMT Relation of Pupils in a primary school.

In the range from seven years old to 13 years old, (7) represents the estimated developmental age of pupils.

$$M = -\frac{\text{NSMT}}{0.0021} + 257 \quad (7)$$

In (7), M is the estimated developmental age by months from birth.

There is a large number of subjects of pupils in a primary school. We can estimate the quadratic approximation of the distribution. (8) is the quadratic approximation of the distribution of all subjects in a primary school.

$$\text{NSMT} = 0.00001145M^2 - 0.004658M + .6781 \quad (8)$$

(8) estimates the distribution of NSMT from 75 months old to 147 months old. If we assume that the tendency represented in (8) continues, the development takes the peak at 17 years old. It is difficult to conclude that the performance of motor control function measured by VST takes its peak at 17 years old. However, this represents the tendency of slowdown of the development of our motor control function.

2) Students in high schools

In Japan, students in a high school are from 15 years old to 18 years old. However, all subjects in this category are 17 or 18 years old. Fig. 6 shows the distribution of all trials. In fig. 6, x-axis represents subject id and y-axis does NSMT. The average NSMT is 0.251.

Using (8), the estimated age of the 0.251 of NSMT is 139 months. Using (7), the estimated age of the 0.251 of NSMT is 137 months. This is same as the sixth grade of a primary school. The total measurements in a primary school and the measurements of 31 high-school students concludes that the motor control function of our brain completes its development at 12 years old. Of course, there are wide personal derivations.

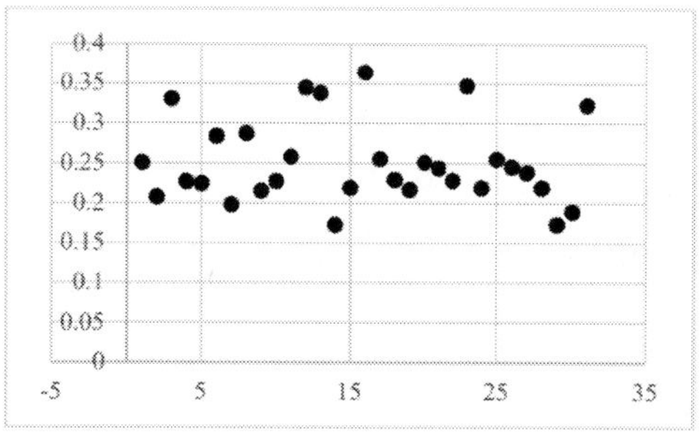

Figure 6. Distribution of NSMTs of High school students.

3) Comparison between Primary Pupils and High School Students

This paper has no subject in a junior high school. The range of ages in junior high school in Japan is from 13 years old to 15 years old. This paper compares the last grade of a primary school and the high school students. In the last grade, we have 66 subjects. However, four measurements have to many errors. Therefore, we reject four measurements, and we have 62 valid measurements in the last grade in a primary school. We have 31 valid measurements of high school students. Table II shows the result of t-test between the pupils in sixth grade of a primary school and the students of high schools. The NSMT of pupils is larger than the one of high schools. However, the probability of both-side in the t-test is 0.26. Therefore, the t-test shows no difference of distributions. The number of subjects in this group is small. We must have much larger scale of measurements of VST.

4) Young people (University Students)

There are 231 trials. However, there are 18 error measurements. As a result, we have 213 valid measurements. Table III summarizes the NSMs at each cycle in young people. At the first cycle, a subject tries to synchronize his hands' motions with the displayed example motion. The average NSM of the first cycle is larger than other cycles. After three cycles, a subject completes the synchronization of his hands to the displayed motions. The NSMs at cycle 3 to cycle 10 are low. At the start of the cycle 11, the example hands image disappears. The NSM at the cycle 11 increases a little. The differences among cycles are small. Fig. 7 shows the average of NSMs in each cycle.

In our experiments, the memory related to simple motion is good in the first five seconds from the disappearance of the proposed example motion shown in Fig. 7. After five seconds, there is a little loss in motion precision.

We computed the difference between the distribution of the NSMs at the cycle 10 and the distribution of other cycles after the cycle 10.

We confirm that they have the same distributions using t-test. Table IV shows the probability of sameness of the distributions from the one of the cycle 10. Fig. 8 shows the probabilities. From cycle 13 to cycle 19, the probabilities are decreasing. This shows that the short-term memory of motor

TABLE II. T-TEST BETWEEN NSMTS OF PRIMARY SCHOOL'S SIXTH GRADE AND HIGH SCHOOL

	6th grade	High school
Average	0.264074945	0.251265719
Distribution	0.002604565	0.002719835
#samples	62	31
Freedom	59	
t	1.124543755	
P(T<=t)	0.132668386	
t	1.671093032	
P(T<=t) Both	0.265336772	
t both	2.000995378	

function decrease rapidly. After cycle 20, the subjects lost the memory about the motion, and their hands' motions became more randomly.

With NSMTs, we estimate the performance of the younger people in synchronizing their hands' movement to the displayed hands' movement. Fig. 9 shows the total distribution of NSMTs in all trials. In Fig. 9, the x-axis represents the trial number. The y-axis is a NSMT. In average, NSMT is 0.217, and the standard derivation is 0.0321. Fig. 10 shows the distribution of NSMTs of younger subjects. The

TABLE III. NSMS OF YOUNG PEOPLE

Cycle	Example motion	Average	Standard derivation
1	Y	0.390	0.153
2	Y	0.266	0.058
3	Y	0.267	0.060
4	Y	0.253	0.054
5	Y	0.256	0.062
6	Y	0.255	0.061
7	Y	0.257	0.103
8	Y	0.252	0.056
9	Y	0.252	0.062
10	Y	0.248	0.059
11	N	0.265	0.070
12	N	0.261	0.074
13	N	0.265	0.078
14	N	0.267	0.068
15	N	0.269	0.076
16	N	0.270	0.083
17	N	0.290	0.135
18	N	0.306	0.185
19	N	0.299	0.151
20	N	0.284	0.147
21	N	0.351	0.771
22	N	0.316	0.355
23	N	0.378	1.086
24	N	0.405	1.468
25	N	0.315	0.156

TABLE IV. PROBABILITIES OF SAMENESS OF NSM TO THE CYCLE 10.

Cycle	Probability of sameness
11	0.0219
12	0.1287
13	0.0305
14	0.0091
15	0.0070
16	0.0060
17	0.0005
18	0.0002
19	0.0001
20	0.0053
21	0.0966
22	0.0194
23	0.1365

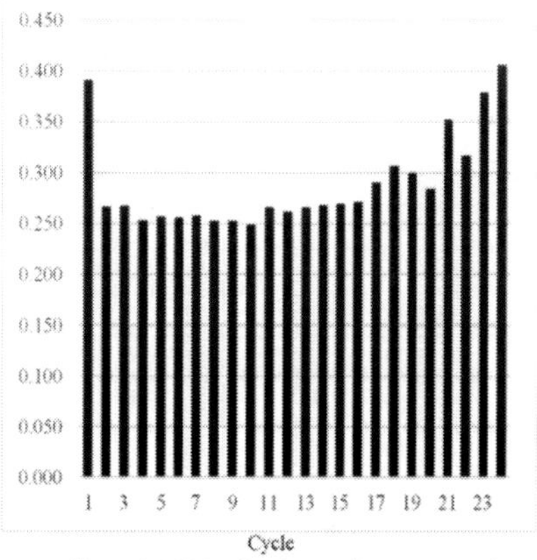

Figure 7. NSMs of each cycle in young people.

NSMTs concentrate around 0.2. Young men show their peak of physical development in twenties [17].

5) Comparison between high school students and young people

Between the distribution of NSMTs of high school students and the one of the young people, t-test shows the result of Table V. The both-side probability is only 0.004. There is a difference between the result of t-test confirms that the distribution of NSMTs of high school students and the one of young people. The performance of brain function estimated from the VST increases from 18 years old to 24 years old. However, the number of subjects of primary school ages are large and nearly complete in a region. The number of subjects of high school ages are small and selected from total ager's assemblage. The subjects of young people are also selected from total ager's assemblage. The young people are students

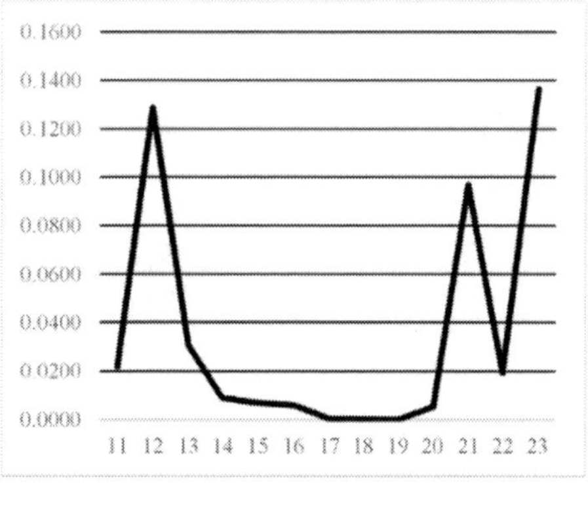

Figure 8. Probabilities of the sameness of the NSMs distributions to the cycle 10.

TABLE V. T-TEST BETWEEN HIGH SCHOOL STUDENTS AND YOUNG PEOPLE.

	High school	Young
Average	0.25684	0.218143
Distribution	0.004508	0.000902
#samples	31	68
Freedom	36	
t	3.071974	
P(T<=t)	0.002018	
t	1.688298	
P(T<=t) Both	0.004036	
t both	2.028094	

of a university. Therefore, the intellectual abilities of our subjects of the young people are better than one of total ager's assemblage.

The difference between high school ages and young people may represent the difference of their brain performance between different populations.

The decrease of NSMT per month is 0.000645 from high school to young people. This is under one-third of the rate of primary school subjects.

In many fields, developments of human functions show growth curve type. The development of brain function measured with VST must show the growth curve also.

6) Comparison between primary school pupil and young people

There is about 10-years difference between sixth grade of a primary school and young people. Table VI shows the result of t-test between 6th grade of a primary school and young people. The t-test confirms that the distribution of 6th grade of a primary school differs from one of the young people.

Using (2), 0.217 of NSMT leads 153.7 months from birth as the estimated age. This is one-year difference from the 6th grade of a primary school. This confirms that the motor control function of our brain slowdown its development over 12 years old.

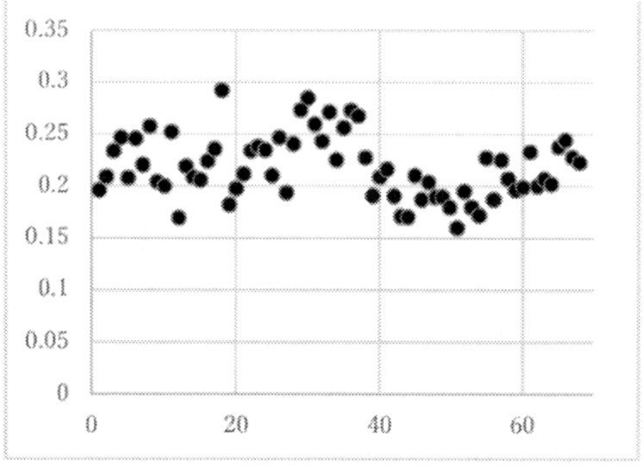

Figure 9. Distribution of NSMTs of Young People.

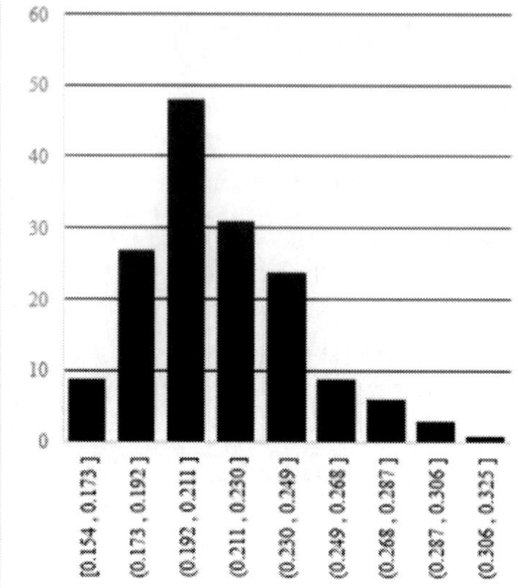

Figure 10. Distribution of NSMTs of young people.

7) Manhood people

Fig. 11 shows the distribution of measurements of all trials. In fig. 11, x-axis is the age of a subject and y-axis is the NSMT. We cannot find the relation between age and NSMTs. The average NSMT is 0.31886. In the distribution, there are some error measurements. Therefore, the author rejects the measurements over 0.6 NSMTs. Over 0.6 NSMT, subjects have some healthy problems. In the case, the average of under 0.6 NSMTs is 0.296. This is also a little worth than the students of high schools. The number of subjects is not large enough to estimate the change of the performance of VST with aging.

With stability in standing posture, we can find the effect of aging after 60 years old [18]. The result of this paper shows same tendency.

8) Elderly people

Table VII summarizes the NSMs at each cycle in elderly people. At the first cycle, an elderly subject synchronizes his hands' motions to the displayed example motion. The average NSM of the first cycle is larger than other cycles. After six cycles, a subject finishes to synchronize his hands to the displayed motions. The NSMs at cycle 4 to cycle 10 are low. At the start of the cycle 11, the displayed example hands image disappears. The NSM increases from cycle 11 to cycle 14. The differences between cycles are not large. Fig. 12 shows the average of NSMs in each cycle.

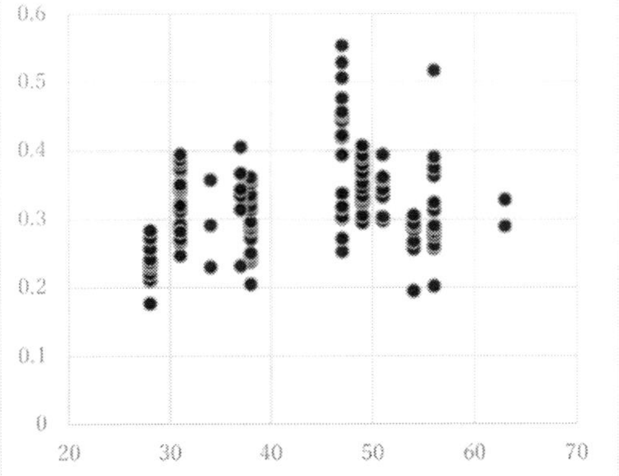

Figure 11. Age-NSMT relation of Manhood people.

Table VI. T-TEST BETWEEN NSMTS OF 6TH GRADE AND YOUNG PEOPLES

	6th grade	Young
Average	0.246935	0.218143
Distribution	0.002862	0.000902
#samples	70	68
Freedom	109	
t	3.912838	
P(T<=t)	7.95E-05	
t	1.658953	
P(T<=t) Both	0.000159	
t both	1.981967	

We have 158 NSMTs of young people and 27 NSMTs of elderly people. On average, the NSMTs of young people are

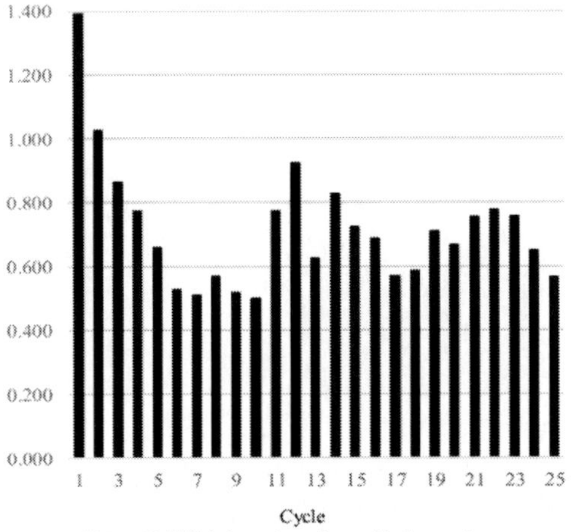

Figure 12. NSMs in each cycle on elderly people.

TABLE VII. NSMS OF ELDERLY PEOPLE

Cycle	Example motion	Average	Standard derivation
1	Y	1.474	0.953
2	Y	1.057	1.212
3	Y	0.887	1.069
4	Y	0.800	1.044
5	Y	0.645	1.016
6	Y	0.490	0.393
7	Y	0.511	0.376
8	Y	0.575	0.654
9	Y	0.438	0.218
10	Y	0.433	0.221
11	N	0.723	0.637
12	N	0.921	1.263
13	N	0.600	0.580
14	N	0.787	0.997
15	N	0.711	0.667
16	N	0.549	0.471
17	N	0.508	0.396
18	N	0.588	0.500
19	N	0.700	0.567
20	N	0.648	0.551
21	N	0.753	0.703
22	N	0.773	0.737
23	N	0.644	0.509
24	N	0.559	0.512
25	N	0.565	0.125

smaller than the NSMTs of elderly people. However, we need to check the reliability. We make t-test with these two groups of NSMTs. Table VIII shows the result of the t-test. The probability of being the same is less than 10^{-9}. The deference between young people and elderly people is significant. This implies that NSMT can measure the deterioration because of the aging process. There is an apparent difference between NSMs of young people and ones of elderly people, as shown in Fig. 7 and Fig. 12.

In elderly people, the deterioration of motor control function increases with the aging process. Fig. 13 shows the relation between the age and the NSMT of each elderly person. The correlation coefficient between the age and the NSMT of

TABLE VIII. T-TEST BETWEEN NSMTS OF YOUNG AND ELDERLY PEOPLE

	Elder	Young
Average	0.395186	0.214459
Distribution	0.009645	0.000892
#samples	27	158
Freedom	27	
t	9.487605	
P(T<=t)	2.17E-10	
t	1.703288	
P(T<=t) Both	4.33E-10	
t both	2.051831	

elderly people is 0.467. There is a linear relation between the age and the NSMT. The linear approximation is (9).

$$\text{NSMT} = 0.0088a - 0.26 \qquad (9)$$

In (9), NSMT is the performance measure of motor control function. a is the years from birth of a subject. The age ranges from 66 years old to 83 years old.

From (9), we have (10) that estimates the age from NSMT.

$$\text{NSMA} = 114\text{NSMT} + 29.5 \qquad (10)$$

In (10), NSMA is an aging years of motor control function. NSMT is a measured NSM at a trial. This shows the measurement of motor control function can estimate the aging of a brain function of elderly people.

The average of NSMTs is about 0.22 in young people. If the deterioration of motor control function shows a linear relation with the years from birth of a subject from the start of the deterioration, we can estimate the motor control function age with (10) over 53 years old people.

The average of NSMTs is about 0.30 in manhood people. If the deterioration of motor control function shows a linear relation. From (10), we have 63.7 years old as the age starting the deterioration with aging process. About 64 years old agrees the starting age of deterioration of aging discussed in [18].

9) Comparison between manhood people and elderly people

The average NSMT of manhood people is 0.296. With (10), the 0.296 of NSMT relates to the estimated age 63.2 years old. Of course, the number of subjects is not enough. The estimation of the starting point of aging is not 63 years old. There are some personal differences of NSMT. Using the normal distribution of NSMTs in manhood people, we have the distribution of starting point of aging. The distribution of manhood people is 0.0632. This leads seven years' distribution from the average 63.2 years old. The effects of aging about VST start from 56 years old to 70 years old. Therefore, we need much more subjects in this range of aging.

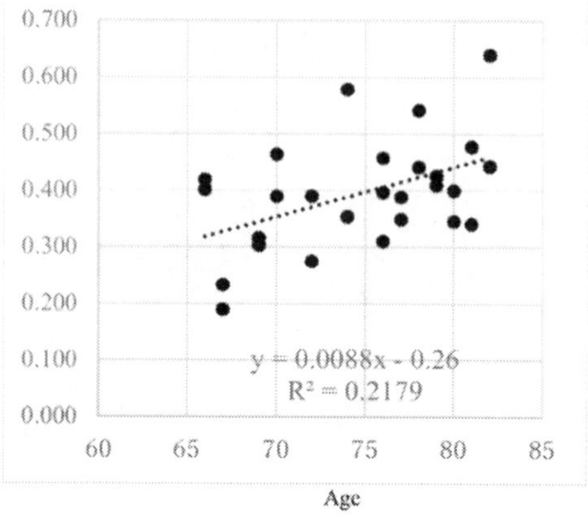

Figure 13. Age-NSMTs relation of elderly people.

10) Total tendency of development and deterioration with aging

From the age of a primary school to elderly people, the outline of the development of motor control function and the deterioration is observed from many experiments shown before. We can summarize the experiments into a total view of development and deterioration of motor control function measured with VST. Fig. 14 shows the total tendency. In Fig. 14, some parts do not relate the ages of subjects. However, they are high school students and young people. High school students are 18 years old. The young peoples are from 22 to 24 years old. The average of the ages of young people is 23 years old.

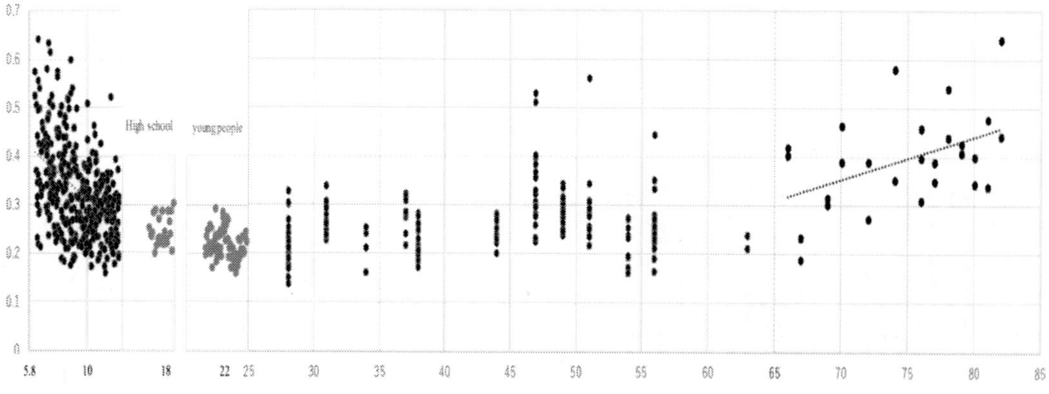

Figure 14. Total tendency of development and deterioration with aging.

In Fig. 14, x-axis is the age of a subject or the average ages of the subjects' group. In the figure, the range of manhood includes small number of subjects. However, the total view of Fig. 14 shows the outline of the development and the deterioration of the performance measured with VST.

Fig. 14 clearly shows the development and the deterioration with aging. Over 65 years old, a subjects show the NSMT under 0.2. These subjects keep their performance of motor control function in their ages.

In Fig. 14, we can see the outline of the development and the deterioration of a motor control function measured by VST. The tendency shown in Fig. 14 agrees the tendency of development in [17] and the tendency of deterioration of aging in [18].

B. Phases

The phase of the measured motion represents the timing of motion. In precise measurements, phases show apparent difference between the first period where the example motion is displayed and the second period where the example is not displayed in a trial. In the part from cycle 1 to cycle 10, the phases keep a similar value. From cycle 11, the phases change gradually. This represents the difference between the speed of the example motion and one of the memorized motions. From this phase change, we can measure the difference of timings between the example motion and the memorized motion.

There is no precise measurement about subjects in a primary school and manhood people.

1) Pupil in Primary school

In a primary school, we have only simple type measurements. In simple type measurement, we cannot have the phase change after the displayed hands' images disappeared. However, we can have the phase stability.

Table IX shows the average phase change between adjacent cycles. Fig. 15 shows the average phase changes of first grade, third grade and sixth grade. In Fig. 15, full line represents the first grade in a primary school. A chain line does the third grade. And, another line does the sixth grade. In Fig. 15, there is an apparent decrease of the average phase changes between adjacent cycles with growth. The phase changes between adjacent cycles decrease with getting older. In Table IX, we can find the development of motor control function. However, phases in VST have large distributions among subjects. Therefore, personal analysis of a subject is difficult.

Fig. 16 shows the average phase changes among grades in a primary school. We can see an apparent progress of motor control function with getting older in Fig. 16.

2) High School Students

High-school students made simple type measurements. Fig. 17 shows the phase changes of high-school students and the sixth grade of a primary school. In Fig. 17, the phase changes of high-school students is less than ones of sixth-grade pupils in a primary school. In averages, phase change of high-school students is 0.227. The average phase change of sixth-grade pupils in a primary school is 0.274. The average phase change of high-school student is 18% smaller than one of the sixth-grade pupils in a primary school. This shows the development of motor control function as NSMTs.

3) Young people

We assume that the phase change in the first part of a trial is smaller than the phase change in the second part of the trial. We divided all cycles into three sections. To confirm this assumption, we calculate the linear approximation of the phases in each section. The first section starts from cycle 4, and ends at cycle10. The second section starts from cycle 11, and ends at cycle 17. The third section starts from cycle 17, and ends at cycle 23, as shown in Fig. 4. In 158 valid trials, there are delay and advance in phases. We evaluate phase change in absolute value.

Table X shows the averages of the slant of each section. The average absolute slant of phases in the first section is smaller than the one in the second section and the third section.

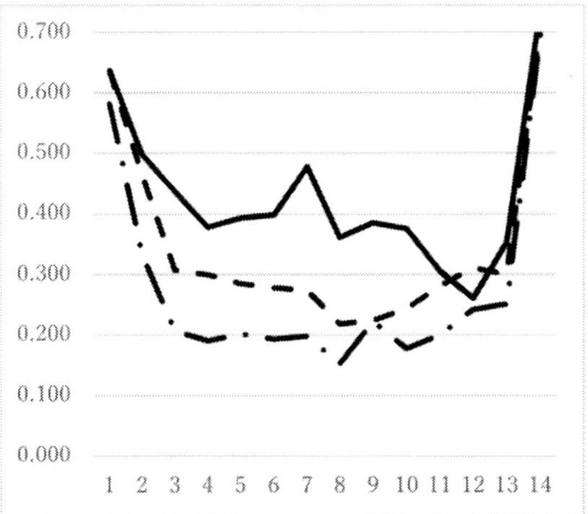

Figure 15. Average phase changes between adjacent cycles.

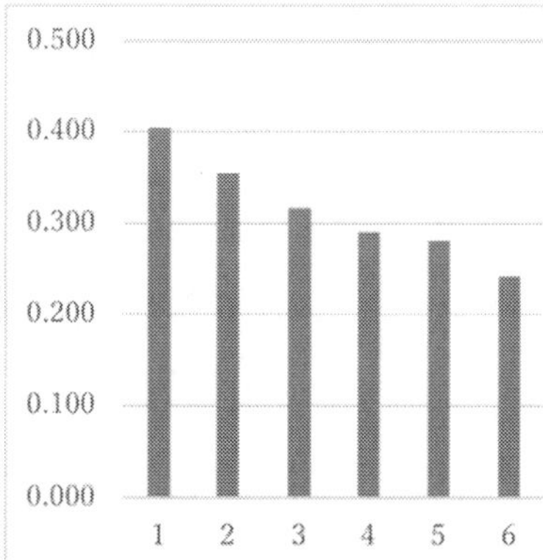

Figure 16. Average phase changes of each grade in a primary school.

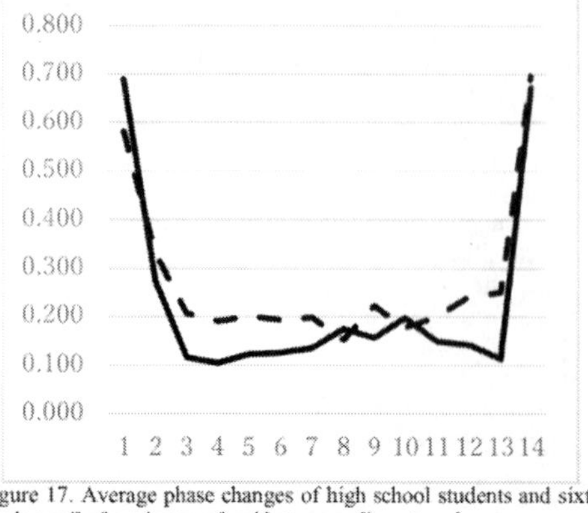

Figure 17. Average phase changes of high school students and sixth grade pupil of a primary school between adjacent cycles.

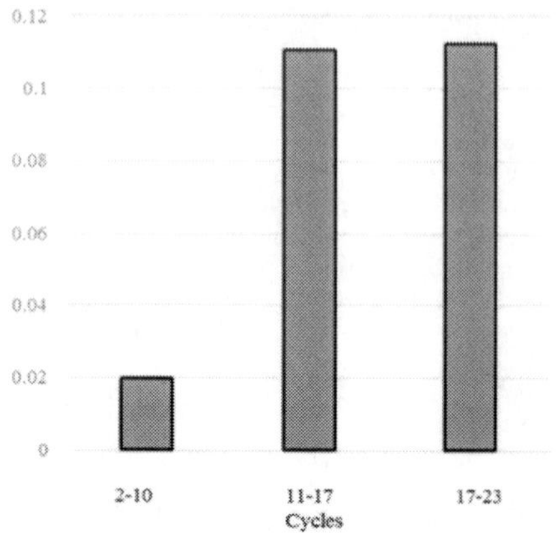

Figure 18. Phase changes in cycles.

Fig. 18 shows this relation. In Fig. 18, there is an apparent increase of phase changes in second and third sections.

Statistically, the first section and the second section have difference bases. Calculation of the t-test confirms that the difference is significant. The t-measure between these two sections is over 12. The probability is under 10^{-26}. The t-measure between the second section and the third section is 0.18. The probability is over 0.85. This confirms that the second and the third sections have a same base. This result means that the memory about the timing of motion remains for at least 15 seconds.

4) Manhood peple

In manhood people, we have only the simple type measurements. Fig. 19 shows the average phase changes between adjacent cycles. In Fig. 19, the full line represents the average of manhood people. The dashed line does one of the high-school students.

The average of manhood people is larger than one of the high-school students. In average, the phase change is 0.272. This is nearly same as the average of sixth-grade pupil of a primary school. These are 20% larger than the average phase change of high-school students.

5) Elderly people

We also calculate the linear approximation of the phases. Table XI shows the average absolute slant of phase's change in each section.

There are apparent differences of the phase changes between young people and elderly people. Elderly people have some difficulties to keep the pace of flipping their hands. In phases, it is difficult to find the proper scale representing an aging process.

C. Discussion

With the NSMs, there is no apparent change between with and without a displayed motion example. Before 15 seconds,

TABLE X. PHASE CHANGES IN SECTIONS OF YOUNG PEOPLE.

Section	Cycles	Slant of phase change
1	4-10	0.022
2	11-17	0.111
3	17-23	0.113

Table IX. AVERAGE PHASE CHANGES BETWEEN ADJACENT CYCLES.

Cycle/ Grade	1	2	3	4	5	6	7	8	9	10	11	12	13	14
1	0.637	0.497	0.437	0.378	0.393	0.398	0.477	0.361	0.385	0.376	0.306	0.261	0.353	0.739
2	0.626	0.572	0.338	0.318	0.356	0.375	0.306	0.330	0.293	0.292	0.240	0.297	0.267	0.679
3	0.635	0.463	0.307	0.300	0.285	0.278	0.273	0.218	0.224	0.243	0.281	0.311	0.300	0.707
4	0.785	0.467	0.274	0.244	0.218	0.225	0.237	0.221	0.182	0.199	0.213	0.291	0.217	0.735
5	0.620	0.473	0.354	0.226	0.232	0.227	0.209	0.217	0.223	0.191	0.214	0.213	0.255	0.748
6	0.582	0.333	0.207	0.190	0.201	0.193	0.198	0.152	0.222	0.177	0.200	0.241	0.250	0.697

there is little decay of the memory of motion. After 16 seconds, Fig. 3 shows a little increase of the NSMs.

With the phases, there is an apparent difference between with and without a displayed motion example. The changes of measured phases represent an error in the timing of a measured motion. Some trials show delay and others show advance. The phase change shows the error about the memory of the timing.

TABLE XI. PHASE CHANGES IN SECTIONS OF ELDERLY PEOPLE.

Section	Cycles	Slant of phase change
1	4-10	0.146
2	11-17	0.323
3	17-23	0.373

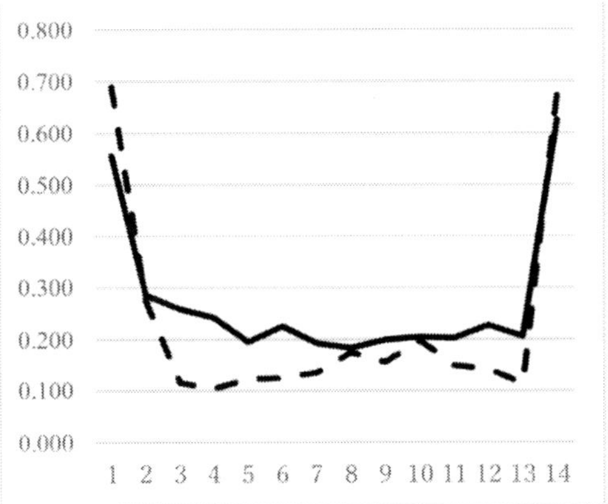

Figure 19. Average phase changes of high school student and Manhood people between adjacent cycles.

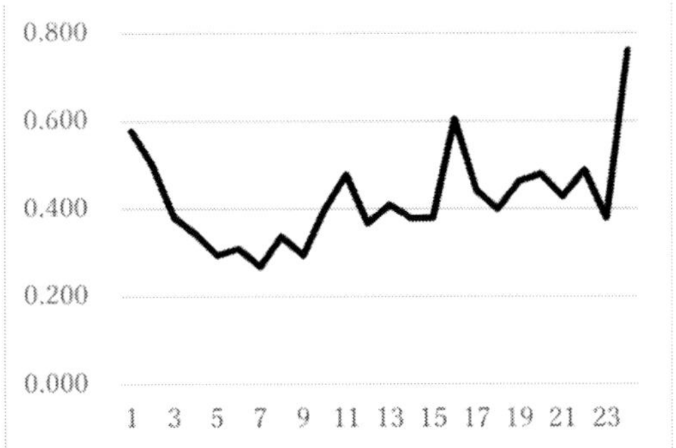

Figure 20. Average phase changes between adjacent cycles of elderly people.

The proposed method measures the timing and the process of movements. A classical tapping test measures the timing only. However, in this experiment, the difference of 0.001 radian in phase is the difference of 0.00016 seconds in time. The proposed timing measure about motor function based on the phase of the basic movement is very keen. The classical tapping test can measure the difference of 0.0001 seconds now. However, the mechanical features about a hand and a switch make it difficult to measure the small difference of time.

Fig. 20 shows the average phase change between adjacent cycles. Elderly peoples need seven cycles to synchronize their hands' movements to the displayed hands' movement. The average phase change between the third cycle and tenth cycle is 0.37. This is about 0.1 larger than one of the manhood people.

Elderly people has much difficulty about synchronizing his hands' movement to the displayed movement. The phase changes between adjacent cycles may have the role to evaluate the performance of motor control function. However, the phase change can have large dependency to the NSM. In the case, we can use only NSM.

V. CONCLUSION

This paper proposes the pair of the measurement and evaluation method of motor control function to estimate whole image of the development and the deterioration with an aging process. The proposed method is implemented and tested in experiments. The proposed visual synchronization task is easy to perform. For instance, it needs only 25 seconds. The proposed Non-Smoothness Measure has enough power of discrimination of a motor control function. The phase changes also have enough power to measure the very small error in timing remembered.

The experimental results confirm that the proposed method can measure and evaluate the development and deterioration of a motor control function with an aging process precisely. This paper proposed the total view of the development and deterioration of the performance of motor control function from six years old to 80 years old, and developed a method to estimate the age according to the aging process of the motor control function in some age groups. In the age group over 65 years old, the estimated age from NSMT helps to measure the deterioration of the brain function, and it can detect the very first stage of cognitive impairment. In the age group in a primary school, the estimated age from NSMT may work for the index of developmental disorder.

Many experiments outlined the development and deterioration of the performance of motor control function with aging process. This helps the understanding of developments and deteriorations of brain function with aging process.

We will perform larger-scale experiments in the next step for more precise understanding of developments and deteriorations with aging process.

ACKNOWLEDGMENT

This work is supported with JSPS16K01057. The authors thank SoftCDC Corporation for their help to many experiments.

REFERENCES

[1] K. Aoki and H. Hotta, "Aging Measurements with Precise Observations of Synchronization Hands' Movements -Aging effects about motor control function-," Global Health 2016, pp. 49-55, October 9 - 13, 2016, Venice, Italy, IARIA, ISBN: 978-1-61208-511-1, 2016.

[2] Lafayette Instrument, "Purdue Pegboard Test," http://www.lafayetteevaluation.com/product_detail.asp?ItemID=159, retrieved 2016/08/29.

[3] S. Hirata, Y. Kitajima, T. Hosobuchi, and M. Kokubun, "The speed and accuracy of fine motor actions in children with intellectual disabilities," Bulletin of Tokyo Gakugei University, 59, pp. 263-267, 2008.

[4] M. Kokubun, "Are children with down syndrome less careful in performing a tray-carrying task than children with other types of mental retardation?" Perceptual and Motor Skills, 88:3c, pp. 1173-1176, 1999.

[5] M. J. Hove, J. R. Iversen, A. Zhang, and B. H. Repp, "Synchronization with competing visual and auditory rhythms:bouncing ball meets metronome," Psychological Research, 77, pp. 388-398, 2013.

[6] Y. Sugano, M. Keetels, and J. Vroomen, "The Build-Up and Transfer of Sensorimotor Temporal Recalibration Measured via a Synchronization Task," Front Psychol. 2012; 3: 246, doi:10.3389/fpsyg.2012.00246, 2012.

[7] V. Krause, B. Pollok, and A. Schnitzler, "Perception in action: The impact of sensory information on sensorimotor synchronization in musicians and non-musicians," Acta Psychologica, Volume 133, Issue 1, pp. 28-37, January 2010.

[8] J. Lee, J. Chai, and P. S. A. Reitsma, "Interactive Control of Avatars Animated with Human Motion Data," ACM Transactions on Graphics (TOG). ACM pp. 491-500, 2002.

[9] S. Vercruysse, J. Spildooren, E. Heremans, J. Vandenbossche, O. Levin, N. Wenderoth, S. P. Swinnen, L. Janssens, W. Vandenberghe, and A. Nieuwboer, "Freezing in Parkinson's disease: a spatiotemporal motor disorder beyond gait," Movement Disorders, 27(2), pp. 254-263, 2012

[10] M. A. Smith, J. Brandt, and R. Shadmehr, "Motor disorder in Huntington's disease begins as a dysfunction in error feedback control," Nature, 403, pp. 544-549, 3 February 2000.

[11] K. Rubia, A. B. Smith, M. J. Brammer, B. Toone, and E. Taylor, "Abnormal Brain Activation During Inhibition and Error Detection in Medication-Naïve Adolescents with ADHD," Am Psychiatric Assoc., 162, 6, pp. 1067-1075, June 2005.

[12] Microsoft, "Kinect for Windows," https://www.microsoft.com/en-us/kinectforwindows/, retrieved 2015/07/07.

[13] Leap Motion, "Leap Motion SDK," https://developer.leapmotion.com/, retrieved at 2015/03/17.

[14] Microsoftstore, "Kinect," http://www.microsoftstore.com/store/msusa/en_US/list/Kinect/categoryID.64752700, retrieved at 2015/07/08.

[15] K. Aoki, and H. Hotta, "Brain Activity Estimation with Precise Motor Measurements of Visual Synchronization Task of Hands," 2015, Global Health 2015, pp.39-44, ISBN: 978-1-61208-424-4, ISSN: 2308-4553, Nice, France, July 2015.

[16] P. Duhamel and M. Vetterli, "Fast Fourier transforms: a tutorial review and a state of the art," Signal Processing, vol. 19, pp. 259-299, Apr. 1990.

[17] Y. Yamada, M. Kimura, E Nakamura, Y. Masuo, and S. Oda, "Limb muscle mass decrease with aging in Japanese men and women aged 15-97 yr," The Japanese Journal of Physical Fitness and Sports Medicine, Vol. 56, 5, pp. 461-472, 2007.

[18] K. Hashizume, H. Ito, H. Maruyama, H. Saito and M. Ishikawa, "Age-Related Changes of Stability in Standing Posture," Journal of the Japan Geriatrics Society, Vol. 23, 1, pp. 85-92, 1986.

Demonstrating and Implementing the Work-Based Access Control Policy for Cooperative Healthcare Environment Using XACML

Mohamed Abomhara and Henrik Nergaard

Department of Information and Communication Technology
University of Agder, Grimsatd, Norway
Email: {mohamed.abomhara, henrik.nergaard}@uia.no

Abstract—**This study focuses on collaborative activities that are best accomplished by organized groups of healthcare practitioners within or among healthcare organizations with the objective of accomplishing a specific task (a case of patient treatment). In our previous work, we proposed an access control model (work-based access control (WBAC)) that is suitable for collaborative healthcare systems in terms of addressing the issues of information sharing and information security. The current study extends on that work by demonstrating and implementing the WBAC access policy for a collaborative healthcare environment to support diverse domains of data authorization management with various constraints. The implementation is based upon using eXtensible Access Control Markup Language (XACML) with SunXACML. We explain the WBAC model for cooperative healthcare systems, introduces a software structure for WBAC implementation, implement the WBAC profile using XACML 2.0, specify permissions and define all authorization policies. Also, we validate the model and compare it with the existing solution to ensure that the model can fulfill and satisfy the main intended objectives. The experimental results demonstrate the efficiency and scalability of WBAC approach. It shows how the WBAC model simplifies decentralized administrative tasks (e.g., changing of team members and shifting responsibilities), thus enhancing the practicability of access control in dynamic collaboration environments.**

Keywords–XACML; Access control; Access control policy; Collaboration environments; Healthcare.

I. Introduction

Information flow concerns how the information should proceed to authorized entities [1], to whom the information should be propagated and what steps and methods should be used to ensure information flow [2]. Secure information flow comprises two related aspects: information confidentiality and information integrity [3]. Information confidentiality involves a set of rules that limits access or places restrictions on certain types of information. Information integrity seeks to prevent an accidental or malicious destruction of information. Different systems have various confidentiality and integrity requirements. For instance, a remote patient monitoring system will have high confidentiality requirements where data must be hidden from unauthorized entities as well as a high integrity checking against random errors due to information sensitivities [4]. Information confidentiality and integrity are increasingly dependent on how the information should flow, to whom the information should be propagated and what steps and methods should be used to ensure information flow.

Access control policies play an important role in ensuring that the information flow is controlled between authorized entities while preserving resource security in the face of inappropriate access [5,6]. Access control policies specify which authorized entities (e.g., user or organization) can perform what operations on specific resources (e.g., files on electronic health records (EHRs) [7,8]). In collaborative environments such as healthcare, it is not easy for traditional authorization mechanisms like role-based access control (RBAC) [9–11] and attribute-based access control (ABAC) [12,13] alone to specify authorization constraints due to the complexity of a continuously growing as well as changing number of users and medical records. In addition to a lack of granularity, manageability and flexibility for the specification and maintenance of policies [14,15].

Moreover, inconsistencies between the access control policies of various individuals or organizations are a common challenge [16]. Due to the dynamic nature of collaboration and team work, it is important to understand to what extent and under what conditions other parties are allowed access rights [17,18]. It is also necessary to employ access control policies to control the way in which information or services are shared between different parties [19,20]. In distributed environments, different participants (individuals or organizations) can play several different roles at a given time (e.g., resource owner, agent or consumer) [1,21,22]. Moreover, each participant manages their own resources and defines their own access control policies. Thus, participants collaborate with each other in various ways, which requires appropriate access control mechanisms in place to ensure that information is accessible only to those authorized to have access [23].

In our previous work [1,21,22,24–26], work-based access control (WBAC) model was proposed. WBAC is extended with the team role concept. A team role classification based on Belbin team role theory [27,28] was proposed [24]. The nine different team roles that Belbin identified were rephrased and classified into *thought*, *action* and *management* [24]. Role is used in conjunction with team role to handle access control in dynamic collaborative environments. Team member must be assigned to one team role (determined by their professional and/or technical knowledge) based on the goal, task and contributes towards achieving the team's objectives. The team role determine the finer role and the extend of access of each team member.

This study extends the previous work [1] to demonstrate and implement WBAC access policy for a collaborative healthcare environment to support diverse domains of data authorization management with various constraints. The implementation is built based on eXtensible Access Control Markup Language (XACML) [29]. The aim is to simplify decentralized administrative tasks and thus enhance the practicability of access control in dynamic collaboration environments.

The remaining parts of this study are structured as follows: Section II presents usage scenarios of collaboration and healthcare data sharing followed by a detailed description of personal role, team role and resource classification. Section III provides an overview of XACML, demonstrates the modeling structures, authorization constraints, request model, policy model, experiments and result. Section IV presents WBAC authorization framework. Validation of the proposed WBAC model and comparison summary with existing solutions are presented in Section V. Discussion, conclusion and future work recommendations are provided in Section VI.

II. Background and Motivation

This section starts with with a short usage scenarios to better understand the collaborations in healthcare domain. This is followed by a description of personnel categories (personal roles, team roles) and the resources classification in WBAC.

A. *Usage scenario: multiple healthcare practitioners cooperation among multiple healthcare organizations*

As shown in Figure. 1, a patient named *Alice* is recently diagnosed with gastric cancer. Surgical removal of the stomach (gastrectomy) is the only curative treatment. For many patients, chemotherapy and radiation therapy are given after surgery to improve the chances of curing. *Alice* entered a cancer-treatment center at her chosen hospital (e.g., hospital A). *Alice* has a primary care doctor (*Dean*) who she regularly visits. Upon entering the hospital, *Alice* also sees an attending doctor (*Bob*) from the hospital. *Alice*'s health condition has caused some complications, so her attending doctor would like to seek expert opinions and consultation regarding *Alice*'s treatment from different hospitals (e.g., hospital B), including *Alice*'s specific primary care doctor who is fully informed about *Alice*'s medical history. Note that the invited practitioners are specialized in different areas, where some are specialists and others are general practitioners. Also, the final medical report of *Alice*'s treatment should be signed by appropriate practitioners using digital signatures [30, 31]. *Alice* should be able to verify the authenticity of the consultation results through the practitioner's digital signature [22, 32].

In such group consultation, also so-called multidisciplinary team consultation [33–35], it is noticeable that:

- Several healthcare professionals are involved in various roles to provide patient care. That includes primary care doctors, general physicians and specialists.
- The care team are formed dynamically and can be readily changed. For example, when *Alice*'s health condition causes some complications, her attending doctor wishes to seek expert opinions and consult with specialists. As a result of a request for a gastroenterology consultation, we assume a gastroenterologist (*Cara*) will join the care team.
- Every participant needs to obtain the medical records they request based on the health insurance portability and accountability act (HIPAA) [36, 37] minimal disclosure principle [38, 39].

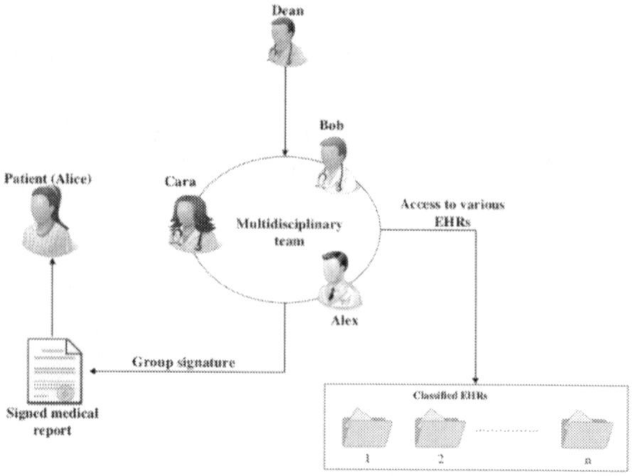

Figure 1. An example scenario of collaboration and sharing of healthcare data [26, p. 4]

- Sharing and accessing healthcare records with efficient coordination between healthcare practitioners is a critical function in access control models [40]. The main concern regards losing control of sensitive healthcare records while sharing them with multiple parties.

The act of managing the collaborative work in a given scenario must be defined clearly. By default, only the main practitioner (*Dean*) should be aware of the patient's personal information. The three other medical practitioners with supporting roles receive information based on their contributing roles based on "minimum necessary" standard to uses and disclosures for treatment [41]. The minimum necessary standard requires covered entities to evaluate their practices and enhance protection of health information as needed to limit unnecessary or inappropriate access to and disclosure of protected health information [41, 42].

B. *Personnel categories: personal role*

A role can be thought of as a set of permissions that a user or set of users can perform within the context of an organization [11, 43]. Permissions are allocated to roles by a system administrator. Such permissions include, for instance, the ability for a doctor to enter a diagnosis, prescribe medication, and add a entry to a record of treatments performed on a patient. Role can be organizational role in which participant has a common set of permissions for performing the job function associated by the name of the role. Example of hospital roles are medical practitioners, nurses and administrators (Figure 2). Moreover, role can be personal roles which represent an individual. They used to create a private workspaces for individuals [18]. Examples of personal roles include pediatric specialists, surgeons or pharmacists. As shown in Figure 3, the role of a pharmacist includes the permissions to dispense but not prescribe prescription drugs.

The role can be statically or dynamically assigned to subject. Static roles are predefined by the organization and manually assigned to users by system administrator, based on a specific organization policy, thereby authorizing users to use the roles' permissions. Membership in a static role is also revoked by a system administrator. The main issues with static roles are how to assigned and revoke them to users and how

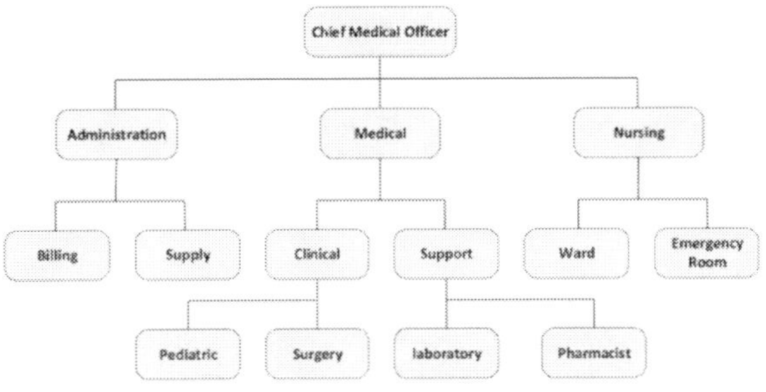

Figure 2. Example of organizational hospital chart

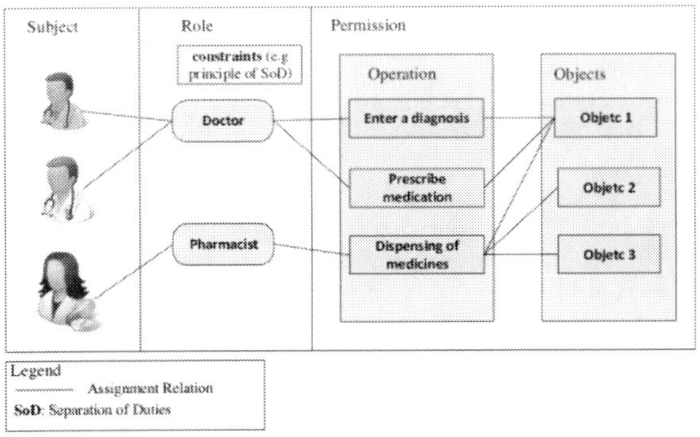

Figure 3. Subject, role and permission relationships

to guarantee that subject are assigned to appropriate roles. An appealing solution is to automatically assign/revoke users to roles. The dynamic role assignment approach has been studied by many researchers. *Al-Kahtani* and *Sandhu* [44] proposed a model to dynamically assign users to roles based on a finite set of rules defined by the organization. Moreover, *Alshehri et al.* [45] proposed a model which uses a concept of pseudorole, which is informally defined as a set of values of static attributes of subject. Although these model tried to solve the problem of assigning users into appropriate roles, it still inherit the major limitation of RBAC, including the lack of granularity and flexibility as well as dynamic adaptability specially in collaborative environments.

The problem of assigning users to role is out of the scope of this study. We assumed that the users within an organization has a role regardless of whether the role has been assigned statically or dynamically. We also believe that, WBAC model can adapt both approaches; static and dynamic subject-role assignments. In our modeling (Section III), we used static role assignment, where we assumed all subject have their roles assigned.

C. Personnel categories: proposed team role

Team is defined as a collection of subjects in specific roles with the objective of accomplishing a specific *work* [46]. Each team has a responsible team manager. Any of the subjects joining a team shares a common goal and may share a default set of permissions for their cooperative work. The notion of a *team role* is used in this study to restrict access permissions to those individuals who not only have the right organizational roles but also are associated to the cooperative *work* via team membership [24].

Regarding the process of collaboration and team work, access control model must be able to provide an efficient and secure platform for people to work together in a hospital without being deterred by restrictive enforcement of access control policies [17]. This can be a rather delicate situation to handle, given the fact that the fluidity of teamwork within the medical domain is often incongruent with technological security. To demonstrate this notion, we consider a scenario (Section II-A) involving four medical practitioners who are working together on a patient's case. For the sake of securing the patient's private (sensitive) data (e.g., mental illness records [47], etc.) [48], the collaboration must be clearly defined. By default, only the main practitioner should be aware of the patient's private information. The three other medical practitioners with supporting roles are given information based on their contributing roles. In order to achieve this, it is imperative to determine the finer roles of each team member. The team role of each member will subsequently determine the extent of access given. The concept of team roles is something that we see as integral to getting the team building process right [33].

Hospital personnel roles are often simplistically split into medical practitioners, nurses and administrators. However, their roles in a team can be further categorized using the team role theory (so-called also Belbin's team roles) [27, 28]. A good collaboration depends on more than team of people working together being enthusiastic and communicating well. Between them, they need the right mix of skills, resources and behaviors to serve the team, too [49, 50]. The Belbin' team role theory is a very useful for higher level team building processes as it helps an experienced facilitator identify the patterns that exist within any team and thus underpin their strengths and weaknesses. Team role theory contains a total of nine roles per group, which are classified into *thought*, *action* and *management* [24] as illustrated in Figure 4.

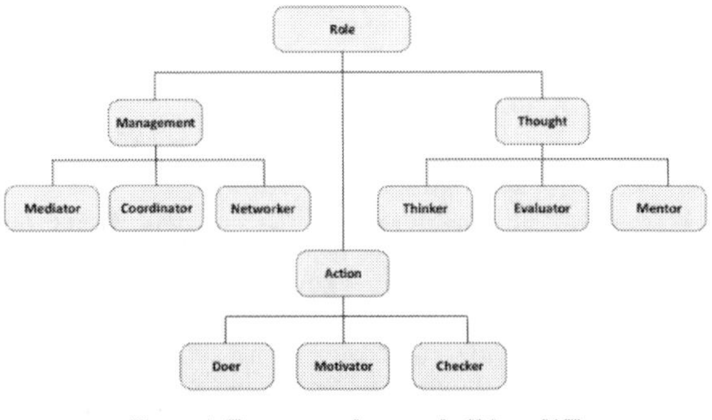

Figure 4. Taxonomy of team role [24, p. 217]

- ***Thought*** denotes a role that is dominated mostly by thinking, analyze problems and/or provide technical expertise. To be a successful thought collaborator, the person may need to understand the medical predicament in detail without necessarily knowing the patient. A worker in this role could be involved in devising strategies to confront particular medical enigmas. Thus, a cardiology specialist may offer his/her expertise regarding the best practices of performing a heart transplant on a child without being involved in the actual operation.

- ***Action***, as the labeling suggests, signifies being involved in task-related collaboration, such as meeting the patient for a medical check-up. Having an action role usually implies close interaction with the patient. Nevertheless, discretion is still feasible with care. For instance, an anesthesiologist needs to only know the patient's physical characteristics to prepare anesthetic. Who the patient is, or where the patient lives is not relevant to completing this task (this assumption is based on our review to [51] (preoperative evaluation and preparation for anesthesia and surgery).
- The ***management*** category comprises personnel who are mostly involved in managing others (e.g., guide, listen, delegate, and solve conflicts). These types of collaborators are adept at coordinating teamwork that is susceptible to social or psychological challenges. For example, in conflict management, they may have to resolve series of opposing diagnoses made by medical practitioners and that may otherwise escalate into serious altercations. In this regard, such personnel's need for information is inwardly oriented. They have a greater need to know personal information about others working at the hospital rather than of patients.

D. Resource classification

Medical records contain a wide range of information, not all of which may be shareable [52]. It could include personal names, phone numbers, addresses, appointment schedules, to do lists, as well as medical history and medical reports regarding patients, to name a few. Some elements of this information may be confidential and sensitive; others may be open for access. In an environment that supports resource sharing, unwanted parties could retrieve the confidential information causing information leakage and leading to the violation of patient privacy. One method to assure that resource sharing will prevent such confidential information leakage is to provide a mechanism to classify all information resources by their degree of share-ability [52, 53].

Medical records classification is infeasible and requires a great deal of effort and skills to accomplish. This is due to issues that, medical records include a variety of documentation of patient's history, diagnostic test results, and daily notes of a patient's progress and medications [54], to name a few. Moreover, healthcare providers can not decide on what appropriate information is really needed for treatment of a patient case. The HIPAA Privacy Rule [37, 55] is a set of standards to protect the privacy of patients' medical records as well as ensure how the health information is used, disclosed and maintained by healthcare organization and health care providers [56, 57]. Healthcare providers should inform and get a patient's permission (e.g., consent or authorization [1]) about how the patient's records are used or disclose? In general, information sharing needed for treatment, therefore, healthcare providers may use and disclose patient records for patient's treatment without a patient's authorization. This could occur during consultation between healthcare providers regarding a patient and referral of a patient by one provider to another. But in most cases when the healthcare providers are dealing with a sensitive information regarding the patient, patient authorization is required for disclosure. For example, The HIPAA Privacy Rule defines psychotherapy notes as "notes recorded by a health care provider who is a mental health professional documenting or analyzing the contents of a conversation during a private counseling session or a group, joint, or family counseling session and that are separate from the rest of the patient's medical record" [47]. Psychotherapy notes are treated differently from other mental health information because they contain particularly sensitive information and because they are the personal notes of the therapist that typically are not required or useful for treatment or health care operations purposes, other than by the mental health professional who created the notes. Therefore, with a few exceptions, healthcare providers must obtain the patient's authorization for any use or disclosure of such an information [47].

Resource within the WBAC is divided into two types, mainly *protected* and *private* resources. *Protected* resources can be shared within a collaborative work. For example, consider scenario (Section II-A), we could say that *protected* object contains resources related to *Alice*'s current case such as past surgical history, date related to abdominal CT scan (computed tomography scan) and gastroscopy data, to name a few. Contrary to the former type, the *private* resources are highly classified pieces of information (e.g., name, data of birth, and address) within the medical records that would be shared during the collaborative work (only if needed). As such, the spreading of access control on the basis of collaboration will not affect the *private* resources. It is meant to safeguard certain confidential information from being leaked out accidentally through collaborative means. In this study we assumed that, personal information (e.g., name, phone number, address, and /or IDs) and any medical records such psychotherapy notes [47] and sexually transmitted diseases (STD) records which are not related to the current medical case are *private* resources.

III. XACML Profile for WBAC

In this section, we demonstrate and implement an WBAC model for a collaborative healthcare environment to support diverse domains of data authorization management with various constraints.

A. An overview of XACML

XACML is a standardized policy language by OASIS [29]. It defines the architecture, policies and messages of an access control system. XACML is a powerful and flexible policy language for heterogeneous distributed systems and is a general-purpose access control policy language [13, 58, 59]. According to the reference XACML architecture shown in Figure 5, the XACML model contains the following main entities [60, 61]:

- *The Policy Enforcement Point (PEP)* is an entity that intercepts a user's request to access a resource. The PEP forwards the request to the PDP to obtain the access decision (i.e., access to the resource is permitted or denied). PEP then acts on the received decision.
- *The Policy Decision Point (PDP)* is used to evaluate access requests against authorization policies and makes decisions according to the information contained in the request before issuing access decisions.

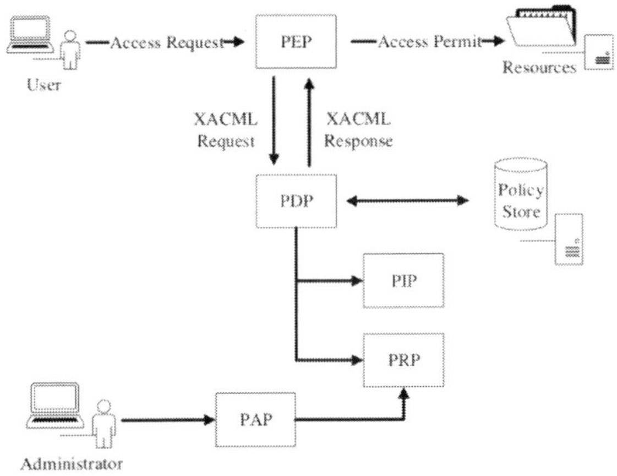

Figure 5. XACML framework

- *The Policy Information Point (PIP)* acts as the source of attribute values, or the data required for policy evaluation (i.e., a resource, subject, environment).
- *The Policy Retrieval Point (PRP)* is an entity that stores the XACML access authorization policies, typically in a database or filesystem.
- *The Policy Administration Point (PAP)* manages the access authorization policies.

The XACML core policy structure (Figure 6) consists of three components: the rule, policy and policy set [61]. The rule is a fundamental component of an XACML policy. The rule, policy and policy set have a target that PDP uses to quickly find the sub-policy parts applicable to making a decision regarding an access request.

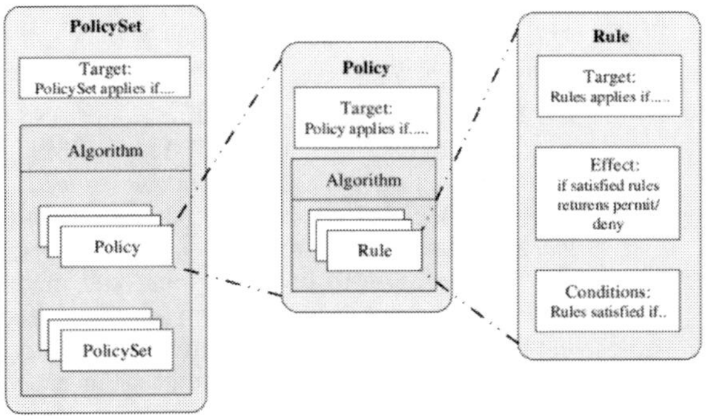

Figure 6. XACML policy structure

The target contains a set of attributes and their values for matching the subject, resource, action and environment, to check if the given rule, policy and policy set are applicable to a specific request. Several rules are grouped and encapsulated into policies and policies are grouped into policy sets. A rule consists of a condition and an effect that can be either a permission or denial associated with the successful evaluation of the rule. A condition represents an expression that refines the applicability of the rule beyond the predicates implied by its target. The correct evaluation of a condition returns the effect of the rule, while incorrect evaluation results in an error (*Indeterminate*) or the discovery that the condition does not apply to the request (*Not Applicable*).

PDP can use different rules, policies and policy sets to make a decision for a specific request. Therefore, conflict might occur between multiple policies when policies offer different authorization decisions. Thus, XACML provides a set of combining algorithms for combining rules and policies to solve a decision conflict between multiple policies [61]. The most commonly utilized combining algorithms are as follows:

1) Deny-overrides algorithm: combines decisions in such a way that if any rule or a policy evaluates denial, then the decision is "deny".
2) Permit-overrides algorithm: combines decisions such that if any rule or a policy evaluates permission, then the decision is "permit".
3) First-applicable algorithm: combines decisions in such a way that the final decision is made based on the first rule or policy in the policy file.
4) Only-one-applicable algorithm: This combining algorithm exists only to combine policy sets and policies. It cannot be used to combine rules. It returns the effect of the unique policy in the policy set that applies to the request; whether Deny or Permit [61].

Based on the combining algorithm used, PDP computes the authorization decision corresponding to the given access request. PDP evaluation is based on the rule, policy and policy set, for which the PDP returns the authorization decision, *Permit*, *Deny*, *NotApplicable* or *Indeterminate*. PDP returns to PEP a sequence of actions called *"obligation"* that should be performed in conjunction with enforcing the authorization decision applied to the access request given.

B. Collaborative work and XACML policy

The *work* model for WBAC (Figure 7) postulates that the entire nature of collaboration can be centralized by the work concept. Here, each *work* is connected to three main components; *personnel* (Section II-B and II-C), *patient* and *resource* (Section II-D). Managing the access control of collaborative work is an interplay between these components.

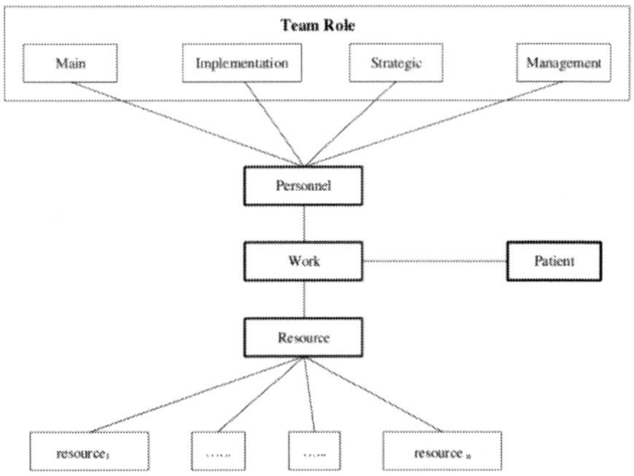

Figure 7. Work model for WBAC

Every resource in WBAC is considered a collaborative entity when it is assigned a *workID*. The *workID* connects the resource to its corresponding work or project that is cooperatively done. By default, a resource does not have a *workID*. This implies that it is not a collaborative resource and thus, cannot be shared. To clarify the idea of managing security through a centralized work, consider the scenario below (Figure 8). Three resources (*resource1*, *resource2* and *resource3*) are all tied to a certain work. As such, all of them contain a *workID* to establish this connection. However, *resource4* is not connected to any work entity. Thus, it does not contain a *workID* and can only be accessed through the main policy.

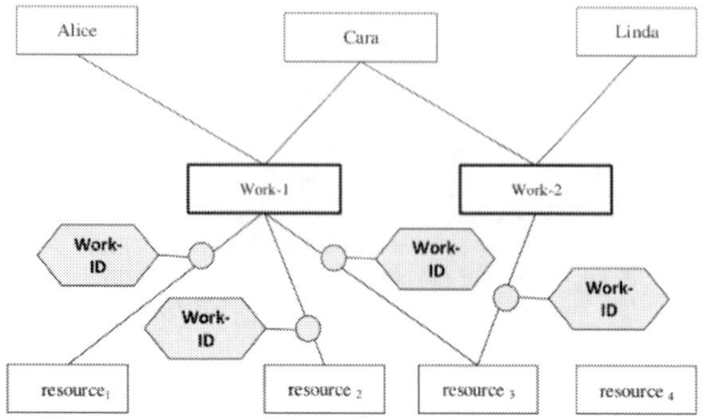

Figure 8. Work and shared resources

Any action that a subject (e.g., healthcare provider) would do on a resource (e.g., patient EHR) is defined entirely within the policy. A dynamic policy with dual inclination is proposed in WBAC [21, 24], whereby the normal policy of enforcing access control is contained within the main policy. On the other hand, any policy that mediates between resource sharing and collaborative work is covered by the collaboration policy. This way, better access control management is achievable. The main policy depends on the roles of the personnel in the organization (e.g., Dean is a general practitioner). PDP only considers the main policy if the personnel possess roles. The collaborative policy is dependent on team roles. In this respect, even if personnel do not have the required roles, they can still gain access upon invitation to collaborate. The team role provides a demarcation between the roles of personnel within a collaboration work and it restricts the role that each team member can have. A person can have various team roles, whereby each is tied to a different collaborative work.

C. Initiation of collaborative work

To begin, the initial situation for access control of which a patient visits the hospital and registers herself. Here, access is given to the physician that she comes in contact with, as well as the nurses at the health institution. As shown in the scenario patient name is *Alice* and her primary care doctor named *Dean*.

In this case of collaborative work shown in case scenario (Section II-A), the workflow of every healthcare practitioner is as follows:

- The primary care doctor (*Dean*) could not solve *Alice*'s case . He invites multidisciplinary team including *Bob*, *Cara* and *Alex* to help. In this team consideration (Figure 9), *Dean* is the core physician of the collaborative work. He serves as the team leader. He is responsible for initiating the work (treatment of *Alice*'s case) and choosing the practitioners (group of doctors) who may be required to attend *Alice*'s consultation and treatment. This implies that he possesses the *main* team role. In other words, he owns the collaborative work initiated. Therefore, full access is given to *Dean* with regard to the information related to the patient. He can access the personal information of the patient as well as the medical records (*private* and *protected* resources). Moreover, the primary care doctor must revoke the group upon completion of the patient's diagnosis consultation.

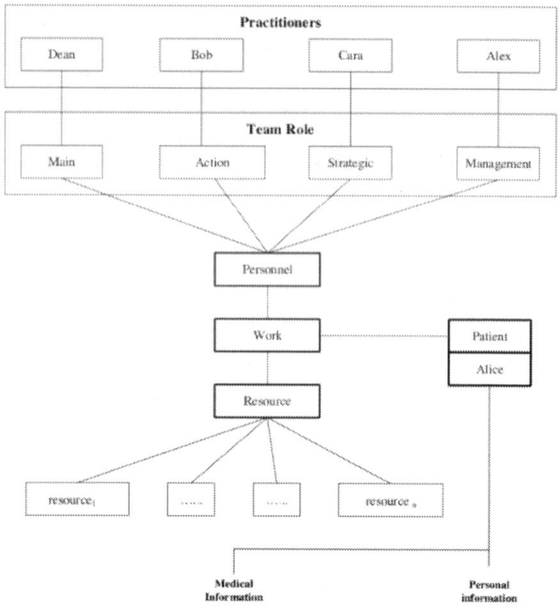

Figure 9. Scenario for team consideration

- *Bob* helps *Dean* with the operational part of the case. Operation refers to a series of responsibilities that entail interaction with the patient. *Bob* needs to see *Alice* on a face-to-face basis to perform various tasks that are related to her recovery. In this respect, there is a need for *Bob* to know personal and medical information about *Alice* to perform his duty effectively. *Bob* is involved in the action part of the collaboration. Therefore, his team role falls under the category of *action*.
- *Cara* has more of a strategy role. She is responsible for helping *Dean* solve the medical case. There is no need for *Cara* to meet *Alice* personally on a day-to-day basis. In fact, *Cara* is only required to analyze the medical situation and suggest a possible solution. *Cara*'s *thought* role within the team implies a rather clear indication of the access that she needs. Since *Cara* is predominantly preoccupied with diagnosing the disease, there is no urgent need for her to know the patient's personal information. As such, she is only given access to the patient's medical information as per her *thought* team role.

- With the increasing number of physicians working on *Alice*'s case, their interaction can become more complex. For instance, if there exists a competition between conflicting diagnoses given by *Bob* and *Cara*, which would gain priority? This is where *Alex* comes in. He contributes to the team by coordinating the interaction of the other members by taking on the team management role. To work effectively, *Alex* does not really need to know the patient's personal information. However, he must be aware of the patient's medical information to enable coordination. Furthermore, *Alex* must also be informed of the work information related to the physicians. In effect, access to certain staff and medical information of the client are given to *Alex*.

In addition, *Alice* may have some historical health information (e.g., mental illness or sexually transmitted diseases (STD), etc.), to which the group (or some of the group) of specialists and practitioners do not have to have access. As we assume in Section II-D that each resource in the system are divided into type, mainly *private* and *protected* during the collaborative work. Each shared resource is tied to the set of collaborative roles or team roles that can access it. In effect, the selected roles will determine the extent of collaborative access.

D. Modeling structures

With the WBAC model, the policy is defined as a tree structure that narrows the combination of attributes presented in an access request. Access to a specific resource is granted when the whole policy tree has found possible matches to the request; the result from rule evaluation is then combined upwards to the outer-most policy using the combining algorithm defined at that level. The result is then sent back to the PEP.

The XACML structure of our model is as follows:

1) Subjects, resources and actions are elements defined by identifier/value pairs (Figure 10). Subjects (e.g., healthcare providers) are entities that send an access request to perform an action (e.g., read or write) on a resource (patient EHRs). The subject is modeled based on the minimal number of attributes required to make different decisions the policy is built to handle. Examples of identifiers are *role*, *employeeID*, *hospitalID* and/or *patientID* (a patient for whom the physician is responsible), to name a few. For the collaborative part, the information about the subject also includes the team identifier for the current collaboration work. As shown in Figure 10(a), physician *Bob* has been assigned the role of attending doctor in the hospital to perform some tasks. He is invited to a collaborative work (*work No 1*) and is assigned the team role *action* to perform some tasks in *Alice*'s treatment.
2) *Collaboration members* comprise a group of healthcare providers (specialists or general practitioners) who are invited to a collaborative work (in our case *Alice*'s treatment). Based on the given scenario, *Dean* is responsible for initiating the work and choosing the practitioners (team of doctors) who may be required to attend *Alice*'s consultation and treatment.
   Bob, *Cara* and *Alex* joined the team and are assigned team roles based on the required job functions. Table I presents the policy data used as an input for XACML. An action represents the operation that a subject can perform on a resource, e.g., *read* and *write* operations. In our model, we also consider several resource attribute as show in Figure 10(b). We also assume the resource are classified into two categories *private* and *protected*.

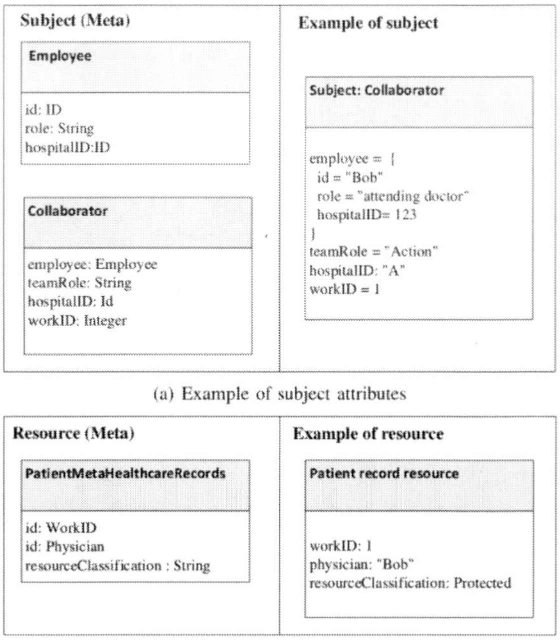

(b) Example of resource attributes

Figure 10. Subjects, resources and actions are elements defined by identifier/value pairs

E. Authorization constraints

We describe the authorization constraints based on our team role classification and our usage-scenario (Section II-A) as follows:

- The subject (healthcare provider) who is assigned the primary doctor role can access both *private* and *protected* resources of the patient for whom he/she is responsible. Figure 11 shows a part of XACML policy ensuring that the primary doctor has a clearance to access medical records.
- A collaborative work must be active, such that team members can work on it. Assuming the value set assigned to a work is its identifier, and if there is no work, the field will not be present in a request.
- Only a subject (healthcare providers) who is a member of the care team and is assigned the *action* team role can access *private* and *protected* resources, but only if needed (inevitably). In this model, we assume the healthcare provider who is assigned the *action* team role needs to access private resources because he/she needs to see a patient on a face-to-face basis to perform various tasks related to the patient's recovery. In this respect, there is a need for the healthcare provider

TABLE I. Tabular structure of policy data

Subject	Job Function	Team Role	Object Type	Action	Permission
Dean	Primary Doctor	Main role	Private and protected	Read/write	Permit
Bob	General practitioner	Action	Private and protected	Read	Permit
Cara	Gastroenterologist	thought	Protected	Read	Permit
Alex	Medical coordinator	Management	Protected	Read	Permit

```
<!--
  Policy ensuring that the primary physician has clearance to access medical records
-->
<Policy PolicyId="team:manager:doctor:record:access:policy" RuleCombiningAlgId="
rule-combining-algorithm:permit- overrides">
  <Target>
   <Subjects>
   <Subject>
     <SubjectMatch  MatchId=" string-equal">
      <AttributeValue   DataType="string">doctor</AttributeValue>
      <SubjectAttributeDesignator  DataType="string" AttributeId="subject:role"/>
     </SubjectMatch>
   </Subject>
   </Subjects>
  </Target>
  <Rule RuleId="isPrimaryDoctor" Effect="Permit">
   <Target/>
   <Condition>
    <Apply  FunctionId=" string-equal">
     <Apply  FunctionId=" string-one-and-only">
      <AttributeSelector
      RequestContextPath="//Resource/ResourceContent/record/patient/physician"
      DataType="string"/>
     </Apply>
     <Apply  FunctionId=" string-one-and-only">
      <SubjectAttributeDesignator  AttributeId="subject:id"  DataType="string"/>
     </Apply>
   </Apply>
   </Condition>
  </Rule>
</Policy>
```

Figure 11. Policy structure for main team role

to know personal and medical information about the patient to perform his/her duty effectively. Figure 12 presents part of XACML policy for *action* team role. It can be seen that a team member who is assigned to the *action* team role (e.g., *Bob*) is allowed access (read only) on both the personal and medical information of the patient (*private* and *protected* resources). Note that in other scenarios, a healthcare provider who is assigned the *action* team role might not need to know private information about the patient.

- Only a subject (healthcare providers) who is a member of the care team and who is assigned to the *thought* team role can access *protected* resources, which are approved for collaboration works. This healthcare provider is predominantly preoccupied with diagnosing the disease, and there is no urgent need for him/her to know the patient's personal information. In fact, he/she is only required to analyze the medical situation and suggest a possible solution. Figure 13 displays a part of XACML policy structure for *thought* team role. In our model (Figure 13), personnel assigned the *thought* team role are permitted access only to *protected* resources (e.g., any resources related to the current case of the patient).
- Healthcare providers who are assigned the *management* team role are responsible for coordinating the other team members' interaction by managing meetings and resolving problems with conflicting diagnoses made by other team members. Figure 14 presents a part of XACML policy structure for *management* team role. The healthcare provider does not really need to know the patient's personal information. However, he/she must be aware of the patient's medical information to enable coordination (Figure 14). Similar to the *thought* team role, personnel assigned the *management* team role are permitted access only to *protected* resources. The difference between the *thought* and *management* team roles is the need for personnel assigned to the *management* team role to have access to team member (healthcare provider) records to be informed of specialist information related to the team members (physicians) in order to coordinate the collaborative work effectively.

```
<Policy PolicyId="actioneer:policy"  RuleCombiningAlgId="
rule-combining-algorithm:permit-overrides">
    <VariableDefinition  VariableId="WorkID">...</VariableDefinition>
 <Target>...</Target>
    <Rule RuleId="permitRead" Effect="Permit">
     <Target>
    <Resources>
      <!--
      Action collaborator shall have access to protected journals of type
    { personalInformation , medicalHistory , patientNote , treatmentSummary }
       -->
       <Resource>
        <ResourceMatch  MatchId=" string-equal">
         <AttributeValue   DataType="string">personalInformation</AttributeValue>
         <AttributeSelector  DataType="string"
         AttributeId="//Resource/ResourceContent/record/type"/>
        </ResourceMatch>
       </Resource>
       <Resource>
        <ResourceMatch  MatchId=" string-equal">
         <AttributeValue   DataType="string">medicalHistory</AttributeValue>
         <AttributeSelector  DataType="string"
         AttributeId="//Resource/ResourceContent/record/type"/>
        </ResourceMatch>
       </Resource>
       <Resource>
        <ResourceMatch  MatchId=" string-equal">
         <AttributeValue   DataType="string">patientNote</AttributeValue>
         <AttributeSelector  DataType="string"
         AttributeId="//Resource/ResourceContent/record/type"/>
        </ResourceMatch>
       </Resource>
       <Resource>
        <ResourceMatch  MatchId=" string-equal">
         <AttributeValue   DataType="string">treatmentSummary</AttributeValue>
         <AttributeSelector  DataType="string"
         AttributeId="//Resource/ResourceContent/record/type"/>
        </ResourceMatch>
       </Resource>
    </Resources>
     <Actions>
      <Action>
        <ActionMatch  MatchId=" string-equal">
         <AttributeValue   DataType="string">read</AttributeValue>
         <ActionAttributeDesignator DataType="string" AttributeId="action-id"/>
        </ActionMatch>
      </Action> </A
        ctions>
     </Target>
      <Condition> <VariableReference  VariableId="WorkID"/>
   </Condition>
</Rule> </Policy>
```

Figure 12. Policy structure for action team role

F. Request model

The XACML request contains the attributes related to subject, resource and action with their corresponding values. For example, in our case and as depicted in Figure 15, we

```
<Policy PolicyId="thought:policy" RuleCombiningAlgId=" rule-
combining-algorithm:permit-overrides">
  <VariableDefinition  VariableId="WorkID">...</VariableDefinition>
  <Target>...</Target>
  <RuleRuleId="protected:resource:rule"Effect="Permit">
    <Target>
   <Resources>
    Thought collaborator  shall have access to protected journals of
                 type: { medicalHistory , treatmentSummary }
      <Resource>
        <ResourceMatch  MatchId=" string-equal">
          <AttributeValue DataType="string">medicalHistory</AttributeValue>
          <AttributeSelector  DataType="string"
          AttributeId="//Resource/ResourceContent/record/type"/>
        </ResourceMatch>
      </Resource>
      <Resource>
        <ResourceMatch  MatchId="string-equal">
          <AttributeValue DataType="string">treatmentSummary</AttributeValue>
          <AttributeSelector  DataType="string"
          AttributeId="//Resource/ResourceContent/record/type"/>
        </ResourceMatch>
      </Resource>
    </Resources>
    <Actions>
      <Action>
        <ActionMatch MatchId="string-equal">
          <AttributeValue DataType="string">read</AttributeValue>
          <ActionAttributeDesignator DataType="string"AttributeId="action-id"/>
        </ActionMatch>
      </Action>
     </Actions>
     <Environments>
     <Environment>
     <EnvironmentMatch MatchId="string-equal">
     <AttributeValue DataType="string"> Hospital.A.Domain </AttributeValue>
     <AttributeSelector DataType="string"
     </EnvironmentMatch>
 </Environment>
 </Environments>
    </Target>
    <Condition>
      <VariableReference  VariableId="WorkID"/>
   </Condition>
  </Rule>
</Policy>
```

Figure 13. Policy structure for thought team role

```
<Policy  PolicyId="Management:policy"  RuleCombiningAlgId=" rule-combining-
algorithm:permit-overrides">
  <VariableDefinition   VariableId="WorkID">...</VariableDefinition>
  <Target>
    <Subjects>...</Subjects>
  </Target>
  <Rule RuleId="managment"Effect="Permit">
    <Target>
    <Resources>
      <!--
Management collaborator shall have access to protected journals of type:
{ medicalHistory ,  treatmentSummary, Doctors information }
      -->
      <Resource>
        <ResourceMatch    MatchId=" string-equal">
          <AttributeValue DataType="string">medicalHistory</AttributeValue>
          <AttributeSelector DataType="string"
          AttributeId="//Resource/ResourceContent/record/type"/>
        </ResourceMatch>
      </Resource>
      <Resource>
        <ResourceMatch    MatchId=" string-equal">
          <AttributeValue DataType="string">treatmentSummary</AttributeValue>
          <AttributeSelector DataType="string"
          AttributeId="//Resource/ResourceContent/record/type"/>
        </ResourceMatch>
      </Resource>
    </Resources>
    <Actions>
      <Action>
        <ActionMatch    MatchId=" string-equal">
          <AttributeValue    DataType="string">read</AttributeValue>
          <ActionAttributeDesignator  DataType="string"AttributeId=" action:action-id"/>
        </ActionMatch>
      </Action>
    </Actions>
    </Target>
   <Condition>
     <VariableReferenceVariableId="WorkID"/>
   </Condition>

  </Rule>
 </Policy>
</PolicySet>
</PolicySet>
```

Figure 14. Policy structure for management team role

have attribute *Subject:Role* and its value *General practitioner*, and attribute *ResourceClassification* and its value *protected* as well as an action value *write*. This information is necessary for authorization decision-making. When PDP evaluates the request against the policy, the attribute names and attribute values are compared according to criteria defined in the policy.

```
<Request>
  <Subject>
    <AttributeAttributeId="subject:id"DataType="string">
      <AttributeValue>Bob</AttributeValue>
    </Attribute>
    <AttributeAttributeId="subject:role"DataType="string">
      <AttributeValue>General  practitioner</AttributeValue>
    </Attribute>
    <Attribute AttributeId="subject:collaboration:work"  DataType="string">
      <AttributeValue>1</AttributeValue>
    </Attribute>
    <Attribute AttributeId="subject:collaboration:role"  DataType="string">
      <AttributeValue>action</AttributeValue>
    </Attribute>
  </Subject>
  <Resource>
    <ResourceContent>
      <record>
        <patient>
        <physician>Dean</physician>
          <work>1</work>
        </patient>
        <classification>protected</classification>
      </record>
    </ResourceContent>
    <AttributeAttributeId="resource-id"DataType="string">
      <AttributeValue>patientRecord</AttributeValue>
    </Attribute>
  </Resource>
  <Action>
    <AttributeAttributeId="action-id"DataType="string">
      <AttributeValue>write</AttributeValue>
    </Attribute>
  </Action>
</Request>
```

Figure 15. Example of an XACML access request

G. *Policies and policy sets model*

The XACML collaboration model begins with a top-level policy set containing one policy for handling a case where the subject is the patient's primary physician and a policy set for the different collaboration cases as shown in Figure 16.

```
PolicySetId="patient-collaboration"PolicyCombiningAlgId="
policy-combining-algorithm:first-applicable">
  <Target>...</Target>
  <!--
    Policy ensuring that the primary physician has clearance to access medical records
  -->
  <Policy PolicyId="team:manager:doctor:record:access:policy" RuleCombiningAlgId="
  rule-combining-algorithm:permit-overrides">...
  </Policy>
  <!--   CollaborationPolicies   -->
  <PolicySet PolicySetId="collaboration:policy:set" PolicyCombiningAlgId="
  policy-combining-algorithm:deny-override">
    <PolicyDefaults>...</PolicyDefaults>
    <Target>...</Target>
    <Policy PolicyId="thought:policy" RuleCombiningAlgId=" rule-
    combining-algorithm:permit-overrides">...</Policy>
    <Policy PolicyId="actioneer:policy" RuleCombiningAlgId=" rule-
    combining-algorithm:permit-overrides">...</Policy>
    <Policy PolicyId="Management:policy" RuleCombiningAlgId=" rule-
    combining-algorithm:permit-overrides">...</Policy>
  </PolicySet>
</PolicySet>
```

Figure 16. Screenshot of top-level policy set

The top-level policy combines the results based on first applicability, meaning that if the requesting subject is the patient's primary doctor, he/she will get access to records regardless of collaboration. PDP will receive all policies as inputs, where each policy has an element known as "target" (described in Section III-A). As depicted in Figure 17, the target element's attribute values (subject, resource, action and environment) are matched with the incoming request (Figure 15) attribute values to decide whether a particular policy is applicable to a given request. If the request attributes match the target's attributes, the policy will be evaluated further. Else, PDP decides the given request is not applicable to the policy.

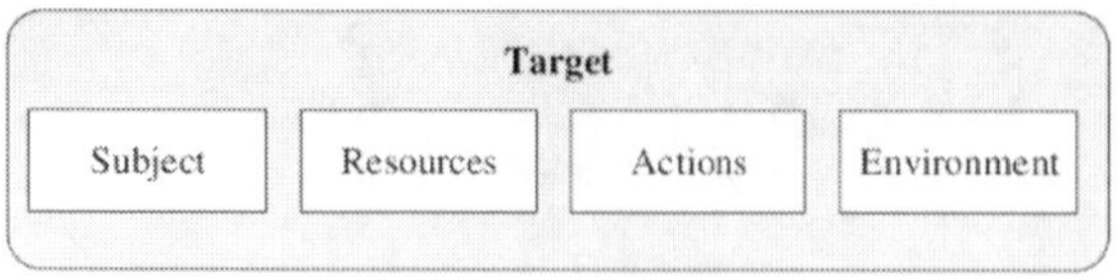

Figure 17. Target element

Within the subject element, XACML uses a sub-element called "subjectMatch" (Figure 18) to define matching criteria for policy. A Subject match element contains two parameters; attribute name and attribute value which are used to compare attribute value with the relevant data type in the policy. XACML engine also uses a sub-element called "SubjectAttributeDesignator" (Figure 18) to look for values from the XACML request related to attribute values from incoming subject (in request). Similarly to "SubjectAttributeDesignator", "ResourceAttributeDesignators" will be used to look for resource in XACML request and "ActionAttributeDesignators" will be used to look for action in the XACML request. The same pattern is applied to "EnvironmentAttributeDesignators".

```
<!--
  Policy ensuring that the primary physician has clearance to access medical records
-->
<Policy PolicyId="team:manager:doctor:record:access:policy"
RuleCombiningAlgId="urn:oasis:names:tc:xacml:1.0:rule-combining-algorithm:permit-overrides">
  <PolicyDefaults>
    <XPathVersion>http://www.w3.org/TR/1999/Rec-xpath-19991116</XPathVersion>
  </PolicyDefaults>
  <Target>
    <Subjects>
      <Subject>
        <SubjectMatch   MatchId="urn:oasis:names:tc:xacml:1.0:function:string-equal">
          <AttributeValue
          DataType="http://www.w3.org/2001/XMLSchema#string">doctor</AttributeValue>
          <SubjectAttributeDesignator DataType="http://www.w3.org/2001/XMLSchema#string"
          AttributeId="subject:role"/>
        </SubjectMatch>
      </Subject>
    </Subjects>
  </Target>
  <Rule RuleId="isPrimaryDoctor" Effect="Permit">
    <Target/>
    <Condition>
      <Apply   FunctionId="urn:oasis:names:tc:xacml:1.0:function:string-equal">
        <Apply   FunctionId="urn:oasis:names:tc:xacml:1.0:function:string-one-and-only">
          <AttributeSelector
          RequestContextPath="//Resource/ResourceContent/record/patient/physician"
          DataType="http://www.w3.org/2001/XMLSchema#string"/>
        </Apply>
        <Apply   FunctionId="urn:oasis:names:tc:xacml:1.0:function:string-one-and-only">
          <SubjectAttributeDesignator AttributeId="subject:id"
          DataType="http://www.w3.org/2001/XMLSchema#string"/>
        </Apply>
      </Apply>
    </Condition>
  </Rule>
</Policy>
```

Figure 18. Sample policy with subject match element

Assuming the subject element in the XACML request (Figure 15), the XACML engine will evaluate the target element by first building an XACML evaluate function contain an *"AttributeId"* and *"AttributeValue"* and de-reference value in "SubjectAttributeDesignator" after matching metadata in policy and XACML request and retrieve appropriate values from request.

The attributes element contains attributes of the entity making the access request. There can be multiple subjects in the form of additional attributes elements with different categories, and each subject can have multiple attributes. In our case (Figure 15), there is only one subject, and the subject has number of attributes. An example of the subject's attribute is subject's identity, expressed as a name ("*Bob*"). Resource element represents the actual resource which subject is trying to access. In the Figure 15 there is an attributes element contains attributes of the resource to which the subject *Bob* has requested access. The resource identified by its classification, which is *protected*. Action element represents subject's activity on a resource (e.g., read and write). An attributes element contains attributes of the action that the subject *Bob* wishes to perform on the resource which is "write". The PDP processing this request context locates the policy in its policy repository. It compares the attributes in the request context with the policy target. The PDP now compares the attributes in the request context with the target of the one rule in this policy.

Figure 11 displays an example of a policy ensuring that the primary physician has clearance to access medical records. While the target element evaluates the applicability of a policy, the rule element implements the actual authorization logic. The primary physician policy has one rule as demonstrated also in Figure 11, which permits access. If the rule's condition is evaluated as true, the output of the rule will be "permit" where the primary physician field in the resource content patient metadata the same identifier for the subject. Condition is a *Boolean* expression (true or false) that refines the applicability of the rule beyond the predicates implied by its target. The effect of rule indicates the outcome of the rule based on the condition evaluation. Two values are allowed: "permit" and "deny".

Collaboration policies are divided into three sub-policy sets from the main policy set, as shown in Figure 16. Each policy set is for one specific team role and the rule that applies to this team role. To evaluate collaborative work, the subject *workID* is matched with that of the resource and must be equal for access to be granted and combined with other constraints, such as *read* or *write* effect. An instance of one collaboration policy is shown in Figures 12, 13 and 14. In figure 13 for example, the subject assigned the *thought* team role is granted access (read access only) to the *protected* resource type if the *workID* matches the active *workID*.

H. Experiments and result

The WBAC model has been implemented using XACML 2.0. Verifying that this implementation of WBAC can be used as part of an XACML policy was done using the Java SunXACML implementation [62] to run a PDP, testing the policy against different requests. Sun's XACML Implementation was originally created in Sun Microsystems Research Laboratories by members of the Internet Security Research Group. It provides complete support for all the mandatory features of XACML and a number of optional features. It also provides support for parsing both policy and request/response documents, determining applicability of policies, and evaluating requests against policies. There are APIs for adding new functionality as needed and writing new retrieval mechanisms for finding things like policies and attributes. All of the standard attribute's types, functions, and combining algorithms are supported [62].

In our experiment, we assume that the PDP is configured to be deny-based which means that any response which is *indeterminate* or *not applicable* is seen as a *deny* response. The WBAC policy was tested by using the attributes based on the data models shown in Figure 10(a) and Figure 10(b) to build access control requests as shown in Figure 15. Both valid and invalid values were set for the different attributes to verify that access was permitted and denied correctly.

The experiments showed that the WBAC model granted access correctly to subjects matching the same work as the resource for the expected cases. Invalid request such as a subject *work* with the value 2, while the resource *work* value set to 1. Since the policy is only implemented with rules needed for permitting access when requests is matched the PDP responded with a *indeterminate* answer, which is interpreted as a deny response when the PDP is deny-based.

IV. WBAC AUTHORIZATION FRAMEWORK

In WBAC, users obtain privileges through roles and team roles. The decision function (PDP) makes a decision for a request permission based on the authorized role and authorized team role. If a role is assigned to a user and is activated, the user will get all permissions associated with the active role. As for team role, the permission a user will get is based on which team he/she is a member of and his/her authorized team role in that team as well as whether the collaborative work is active or not. As shown in the request model (Figure 15), the request should contain all information (attributes) about the user, operation and object including the user's authorized role and authorized team role.

WBAC enables determining if the user, once identified, is permitted to access the resource. According to Figure 19, WBAC is a combination of authentication and authorization processes aimed at managing and securing access to system resources while also protecting resource confidentiality and integrity, among others.

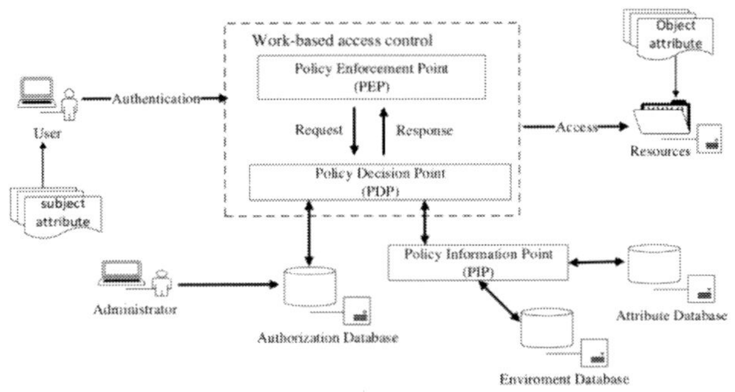

Figure 19. Access control mechanism for WBAC [32, p. 185]

Authentication entails validating the identity establishment between two communicating parties, showing what or who the user is. Authorization checks if the user can access the resources he/she has requested. When a user requests access to a system resource, the user must first authenticate him/herself to the system. In our work [22, 32], we proposed an attribute-based authentication (ABA) scheme, which is a way to authenticate users by attributes or their properties. Second, the WBAC authorization process decides to permit or deny the access request based on the authorization policies. PEP intercepts a user's request to access an object and then forwards the request to PDP to obtain the access decision (permit or deny). PDP receives the request from PEP and combines the user with the object information (attribute value described in Section III), then checks if they satisfy the authorization policies (Figure 19). If so, the subject's access request is granted and will be enforced by PEP.

A. *Evaluation process and decision-making*

Figure 20 presents a sequence diagram of the authorization evaluation process for the WBAC model. When a user sends an access request (Figure 15) to perform an operation on an object, PEP intercepts the call request and forwards it to PDP (access decision function) to check whether the user has permission to perform the requested operation on the object. The authorization system decides if the user has permission to carry out the requested operation by checking three layers: the first RBAC layer, the secondary RBAC layer and the ABAC layer.

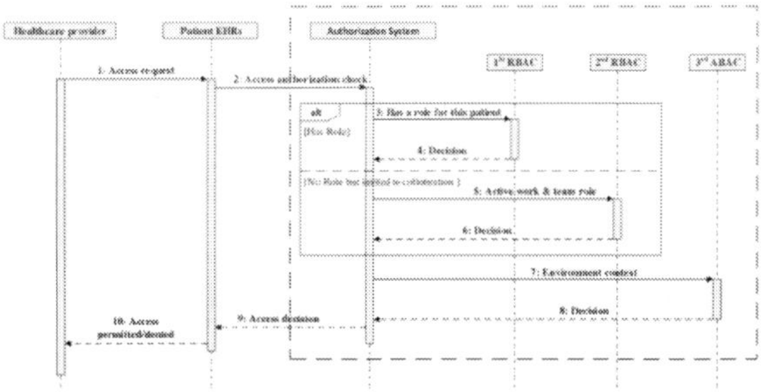

Figure 20. Sequence diagram of authorization process

The entire authorization process is shown in Figure 21. The authorization system is responsible for making an authorization decision on an access request by checking if the access request should be permitted or denied. The access checking operation starts with gathering all attribute values in the access request (e.g., user role, object, and operation attributes) followed by checking the user's state – whether the user is in the user set. If the user is active, the checking process continues with a role check, team role check, and permission check, otherwise the checking process stops and returns the value "no".

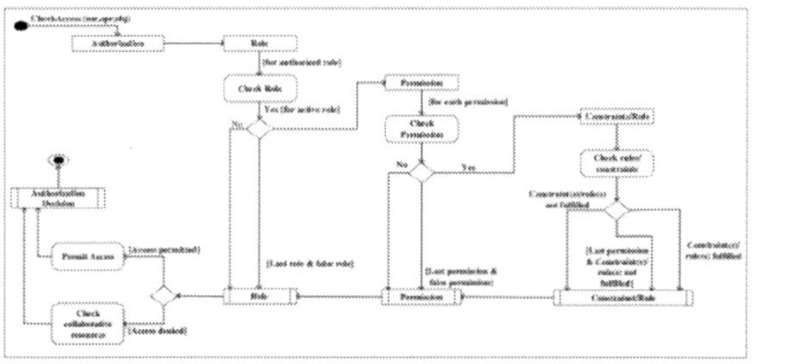

(a) Activity diagram of the role check authorization process

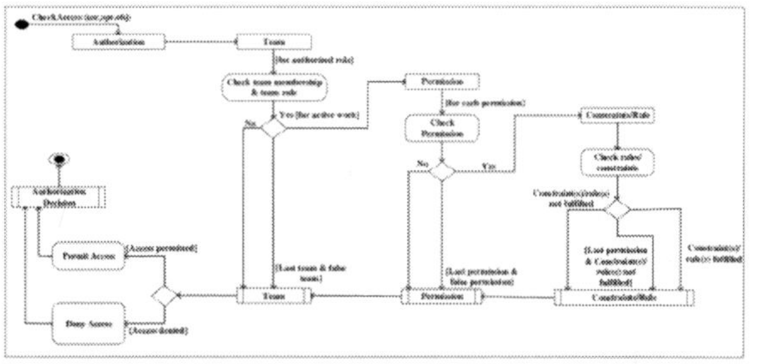

(b) Activity diagram of the team role check authorization process

Figure 21. Activity diagrams of the WBAC authorization process

The role check process (Figure 21(a)) performs a role lookup to check if the role is assigned to the respective user. Only when the user is assigned the role the check process continues with the permission check, otherwise it stops and returns the value "no" and the check access operation investigates the collaborative resources (Figure 21(b)). The permission lookup process checks whether the requested operation on the respective object is assigned to the corresponding role and if the input request object is equal to the permission object. If the requested operation is permitted by the role, the check access operation return "yes" and continues with the constraint and rule check on the ABAC layer. To provide a fine grained access control, the third layer (ABAC) enforces extra constraints such as environment and context constraints. It is not sufficient to grant access only when the user holds the appropriate role.

In case the permission in the request is not assigned any role or the constraint check returns "no", the check access operation further investigates the collocation policy (Figure 21(b)). The check access operation checks the user memberships in a team and if permission is granted by the team role. If the request is permitted by the respective team role and the input "request object" is equal to a permission object, the check access operation returns "yes" and continues with the constraint and rule check on the ABAC layer; otherwise the check process stops and the access request is denied.

Consider *Alice*'s case presented in Section II-A with four healthcare providers: *Dean*, *Bob*, *Cara*, *Alex*. If *Dean* sends a request to read *Alice*'s file in *Alice*'s *private* objects, the check access operation checks if the permission (e.g., read *Alice*'s private object) is assigned to *Dean*'s role (primary doctor). Based on the our defined policy (Table I), *Dean* is assigned the primary doctor role and the permission (read *Alice*'s private object) is assigned to the primary doctor role. Therefore, based on the role and permission checks, *Dean* is permitted to perform the operation "read" on *Alice*'s *private* objects. However, granted access based on an appropriate role is not sufficient. Thus, WBAC facilitates more fine-grained access by checking the third layer (ABAC) for additional constraints, for example if *Dean* is permitted to read a file form a certain location at a particular time. In *Dean*'s case, the authorization system checks only the main policy set, where the requesting subject is the patient's primary doctor.

If *Bob* sends a request to access *Alice*'s EHRs, the access policy (Table I) shows that *Bob* is assigned a general practitioner role, but based on the permission check, permission (read *Alice*'s private object) for example is not assigned to the general practitioner role; hence, the permission check returns "no" and the check access operation continues checking the collaboration resources (Figure 21(b)). In our model, it is assumed that Bob joined *Alice*'s treatment team and is assigned an *action* team role. Therefore, *Bob* is a member of the team and holds an *action* team role. The team check returns "yes" and the check access operation continues with permission checking. Permission (read *Alice*'s private object) is assigned to the *action* team role, thus *Bob* is permitted to read *Alice*'s *private* objects.

V. VALIDATION OF THE PROPOSED WBAC MODEL

In this section, we present a validation of WBAC to ensure that WBAC strikes a balance between collaboration and safeguarding sensitive patient information.

A. *Informal Validation of WBAC*

This informal validation examines the core functions of access control models [10]. The core function as following:

- **Initiation of collaborative work**: the process of initiating the collaborative work (discussed in Section III-C).
- **Policy structure**: Policy is a statement of what is, and what is not allowed and policy structure is a procedure for enforcing the policy in the system (discussed in Section III-G).
- **Alteration of policy for collaborative work**: the process of altering access control policies by WBAC model to meet the requirements of the organization and collaborative work. Consider again the case of *Alice* of which *Cara* plays a *thought* team role in deciding the best treatment for *Alice*'s case. Since *Cara* does not need to see the patient on a face to face basis, she often contemplates upon the decision making from her local hospital (we assumed that *Care* is invited from hospital B). This implies that the shared resources for the *thought* team role are not accessed at the hospital. The alteration that enables the aforementioned scenario can be seen below (Figure 22). Observe that the first rule (Figure 13) allows anyone from the *thought* team role to read the shared information locally (e.g., in hospital A). On the other hand, the second rule alters the former policy. The physician with *thought* team role can access the shared resources from other location (e.g., in hospital B). In both cases however, only the read access is given. Also, the modification will be done in the collaboration policy set. There is no need to modify any policy in the main policy set.

```
<Policy PolicyId="thought:policy" RuleCombiningAlgId="rule-
combining-algorithm:permit-overrides">
 <PolicyDefaults>...</PolicyDefaults>
 <VariableDefinition VariableId="WorkID">...</VariableDefinition>
 <Target>...</Target>
 <Rule RuleId="protected:resource:rule" Effect="Permit">
  <Target>
   <Resources>...</Resources>
   <Actions>...</Actions>
   <Environments>
   <Environment>
   <EnvironmentMatch MatchId="string-equal">
   <AttributeValue DataType="string"> Hospital.A.Domain </AttributeValue>
   <AttributeSelector DataType="string"
   </EnvironmentMatch>
   </Environment>
   <Environment>
   <EnvironmentMatch MatchId="string-equal">
   <AttributeValue DataType="string"> Hospital.B.Domain </AttributeValue>
   <AttributeSelector DataType="string"
   </EnvironmentMatch>
   </Environment>
   </Environments>
  </Target>
  <Condition>...</Condition>
 </Rule>
</Policy>
```

Figure 22. Policy structure that involves alteration

- **Alteration of permission for collaborative work**: the process of altering assigned permissions to subjects to access an resource. The permission of accessing the resources that are related to the collaborative work is reliant on the given team roles. This could change dynamically. For instance, supposed that *Dean* is answering a compelling medical emergency call that forces him to leave the country. To ensure the fluidity of the collaborative work, he promotes *Bob* as the *main* team role. With the new team role, *Bob* is given much greater control over the collaboration. Therefore, there is a need to change the permission to reflect the new role more accurately. This is simply done by altering the *action* team role that was initially defined for *Bob* to the *main* team role as shown below (Figure 23). The change only affects this particular collaborative work and nothing else.

Personnel	Team Role
Bob	Action

⇓

Personnel	Team Role
Bob	Main

Figure 23. Alteration of permission

- **Termination of collaborative work**: the process of deleting all assigned permissions to collaborative Work. Via successful collaboration, the right diagnosis for *Alice* is obtained. After receiving the required treatment, *Alice* is now fully recovered and left the hospital. The collaboration between *Dean*, *Bob*, *Cara* and *Alex* is no longer needed. Subsequently, *Dean* completes the final report for *Alice* and withdraws the collaborative work. Now, supposed that in the future, if *Bob* is inclined to review the diagnosis, then he must request for access again. When the owner of the collaborative work deletes or withdraws the project at hand (Figure 24), all the access to the shared resources, including those that contain the medical or personal information of the patient are revoked. The *workID* that is tied to their access is therefore deleted. Deletion may entail an exhaustive search by the system to guarantee complete removal of access to shared parties. In effect, the other collaborators will cease to have access over the information related to the work. A timestamped log entry of when a work participant entered the work flow should be made, and a corresponding timestamp of when the *work* was completed (which is when the work rights were revoked).

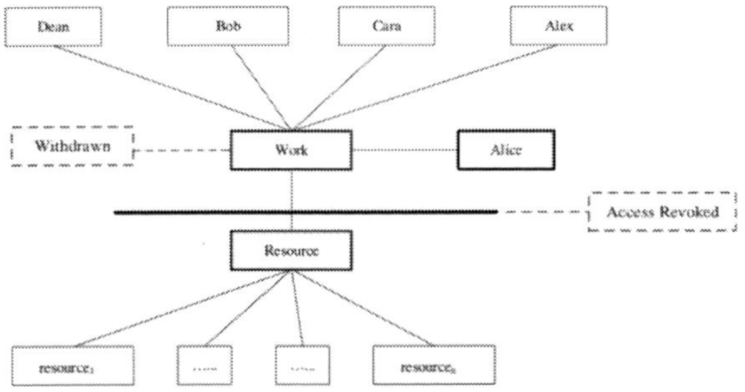

Figure 24. Work withdrawn to terminate collaboration

B. Comparing WBAC with the existing solutions

Researchers have made the best effort to propose an access control model that balance between security and collaboration requirements [23, 63, 64]. A numerous of research trends on access control approaches have been presented such as RBAC, ABAC, team-based access control (TMAC) [46], task-based access control (TBAC) [65], context-based TMAC (C-TMAC) [66], team task based RBAC (TT-RBAC) [67] and group-based RBAC (GB-RBAC) [68]. In this section, we compare them to understand better the differences between these approaches. Comparison is imperative and aims at well defining the appropriate access control model for our model. The main evaluation criteria for access control in collaborative system were presented in number of studies [45, 64]. The assessment criteria with respect to healthcare collaborative environments as follows:

1) **Personalized permission**: Patients must be informed of the collaboration and should be given the right to choose who can have access to their records.
2) **Selective confidentiality**: Certain patient information is highly sensitive. Thus, patients should be able to withhold information that remains confidential.
3) **Flexibility and adaptability**: Flexibility is the access control model's ability to support frequent changes in policy, whereas adaptability is used to evaluate the access control's ability to adapt to different healthcare scenarios and environments.
4) **Fine-grained control**: The access control model should support fine-grained subjects, objects and access rights. This is a granular level at which rules can be applied not only to roles but also to individuals regarding one or many controlled objects [64].
5) **Groups of users: assignment and revocation**: in collaborative work, common tasks are undertaken by a group of people (a team). Therefore, an access control model supports the team's notion and facilitates specifying access rights for teams. Also, the model should have the capability to revoke subjects' access rights to objects.
6) **Policy specifications and maintenance**: The access control model should allow for scalability and easy extension and modification of subjects' access rights to objects. Also, it should provide means of ensuring correct enforcement of the policy or constraint specification.
7) **Design for collaborative healthcare systems**: This criterion indicates whether the access control solution was designed specifically for collaborative healthcare systems.

Table II summarizes our comparative analysis of the RBAC, ABAC, TMAC, TBAC, C-TMAC, TT-RBAC, GB-RBAC and WBAC models. It can been seen that WBAC meets the requirements of collaborative healthcare environments better than the other models. The WBAC model solves the problems of personalized permission and selective confidentiality, whereby, as described above, access to objects is controlled based on the classification of teams into three classes according to the team members' tasks they will carry out in the collaborative work. RBAC, TT-RBAC, GB-RBAC, TMAC and other models do not consider team classification. They deal with all teams in the same way, which can confuse

TABLE II. Comparative analysis of the RBAC, ABAC, TMAC, TBAC, C-TMAC, TT-RBAC, GB-RBAC and WBAC models

Access Control models	Assessment Criteria						
	1	2	3	4	5	6	7
MAC	No	High	Low	High	No	Complex	No
DAC	Yes	Low	Low	Low	No	Complex	No
RBAC	No	Low	Medium	Low	Yes	Simple	Yes
ABCA	No	Yes	High	High	Complex	Complex	No
TMAC	No	Low	High	Yes	Yes	Simple	No
TBAC	No	Medium	Low	High	No	Complex	No
C-TMAC	No	Low	High	Yes	Yes	Complex	No
TT-RBAC	No	Medium	High	Medium	Yes	Complex	Yes
GB-RBAC	No	High	Medium	Low	Yes	Complex	No
WBAC	Yes	High	High	High	Yes	Simple	Yes

security administrators and object owners. In WBAC, the patient will be informed about the team formation and to what information each team member will get access based on the assigned team role. WBAC supports selective confidentiality well because it is possible to assign a specific object to each member in a given team based on the object and team role classifications.

Considering TBAC and TT-RBAC, tasks in healthcare environments usually have their own (different) characteristics and it is difficult to establish in advance access based on tasks. For instance in *Alice*'s case, it is hard to identify what task *Bob* has. In the WBAC model, as *Bob* is assigned the *action* team role, he would have all tasks related to preparing *Alice* for operation. Examples of *Bob*'s tasks are laboratory work (e.g., taking all blood tests required for the operation) and physical examination (e.g., physical examination based on gathered information related to past and current medical history, surgical history, family history, social history (use of tobacco, alcohol and illegal drugs), history of allergies, and current and recent drug therapy [51] to name a few). *Cara* is assigned the *thought* team role. Therefore, her tasks might be for example preoperative risk assessment (e.g., function of the patient's preoperative medical condition) and treatment recommendations after surgery (e.g., pain management post-op [69]). In these cases, access privileges are assigned to healthcare providers according to their team roles and not their tasks. Holding a team role would allow healthcare providers to access multiple information (based on the selective confidentiality requirement), which would allow them to work on multiple tasks related to the patient's treatment. Thus, healthcare providers assigned to the team would be permitted to access the selected objects required for performing their duties.

In terms of fine-grained control, WBAC focuses on the user's role, user's team roles and target object; therefore, it can be said WBAC is classified as fine-grained access control. WBAC reduces over-privilege access arising from frequent specifications when using role in RBAC by classifying the team and objects. The level of fine-grained control access (granularity) to objects that can be authorized to healthcare providers is managed and controlled based on individual scenarios (active work, which is the patient's treatment). Although fine-grained control is very complicated in healthcare environments, WBAC's policy can be implemented using XACML, and the more information that is considered to define a rule, the finer-grained the resulting access control will be. XACML can specify rules in terms of attribute values (e.g., attributes about users, resources, actions, and the environment) that can be of various types, such as strings and integers (Section III-D), making WABC very fine-grained.

WBAC supports an easy means of adding, changing, manipulating, and specifying a team of users. Regarding groups of users, assignment and revocation are similar to TMAC, C-TMAC and TT-RBAC, except that in WBAC the team is classified based on team role. Moreover, in WBAC, a team can be assigned to a collaborative work at any granularity based on the team members' team roles. In general and as explained in [70], using the concept of role in RBAC and its extension greatly reduce the management complexity of user assignment and revocation. Thus, employing the team role concept in WBAC helps solve the problem of user assignment and revocation in the case of team work.

Policy specification and policy enforcement in WBAC are the same as in RBAC. WBAC supports means of specifying and managing policies as well as using appropriate policy languages such as XACML (Section III), which allows extensions or modifications in a simple and transparent manner. The proposed dual policy [24] is to ensure system scalability, especially in collaborative environments, where governance policies require different organizational entities to have different responsibilities for administering various aspects of policies and their dependent attributes.

WBAC has a number of advantages including flexibility in terms of permission administration management, since roles and team roles can be updated without updating permissions for every user. Moreover, it is fairly easy to assign and revoke users based on their roles and team roles. We believe that WBAC handles personalized permissions well and meets our expectation of allowing fine-grained access control, and it enhances the practicability and manageability of access control in dynamic collaboration environments.

VI. Discussion and Conclusions

In this section, discussions, conclusions and future works are presented.

A. Discussion

To prevent any violation of the access control policy of an organization, most classical access control models like RBAC and ABAC define users rights precisely, based on subject and object elements. When several subjects and objects are involved, the subject-object model cannot deliver satisfactory security management. In collaborative environments such as healthcare, it is challenging to predefine all access needs based on the subject-object model. One example of such a situation is explained in our case scenario (Section II-A), which may not be predictable and it would be hard to express the condition of who should join the collaboration and when *Dean* necessitates collaborative support from other parties. Moreover, in deciding on the extent and limit of resource sharing, For instance, in the case of *Alice*'s treatment, which sensitive data should be disclosed to an assisting practitioner so collaboration can be effective, and which should be hidden to safeguard the patient's privacy? Another important matter is the correctness of the policy. Access policy adoption may be limited if the intended policies are not implemented efficiently and consequently thus perform poorly.

WBAC was proposed to address these concerns and support the security and collaboration requirements in access control [23, 63, 64]. The major contributions of the WBAC model include ensuring that access rights are dynamically adapted to the actual needs of healthcare providers and providing fine-grained control of access rights with the least privilege principle, whereby healthcare providers are granted minimal access rights to carry out their duties. In our case scenario, it was noted that general practitioner *Dean* could not solve Alice's case alone. He invited a multidisciplinary team including *Bob*, *Cara* and *Alex* to help. In this team, *Dean* is the core physician in the collaborative work and servers as the group manager. He is responsible for initiating the *work* (*Alice*'s treatment case) and choosing practitioners (group of doctors) who may be required to attend *Alice*'s consultation and treatment. This implies that *Dean* holds the main role. In other words, he owns the initiated collaborative work. Therefore, *Dean* is given a full access (based on his role as primary physician, Figure 11) with regard to patient-related information. *Bob*, *Cara* and *Alex* are assigned to team roles based on the job function they will perform in *Alice*'s treatment. In our previous work [21], we formally describe and showed how each user joins the team and how each should be assigned at least one team role; a team role can be assigned to none or multiple users in many teams.

In this study, we showed how XACML can be used to implement the WBAC model policy and how XACML combining algorithms can be used to manage the inconsistencies between different policy sets. We selected XACML because it has been proven to be adaptable to specifying several common access control methods, such as RBAC and ABAC. Moreover, XACML has become very popular in both academia and industry as a standard for combining, maintaining and exchanging access control policies. It is an architecture for evaluating authorization requests and for issuing authorization decisions. The experiments we conducted demonstrated the applicability of XACML to supporting collaborative and distributed domains in sharing access control of specific resources. However, It still come with some limitation in the expressive power of higher-order logic such as the expressions of separation of duty (SoD) constraints and domain constraints.

Our implementation only covers access request for medical records resources, but by using similar matching technique as for the *work* attribute, it is possible to extend this to other polices that are also active during collaboration. An example of this could be for persons with the *management* team role, which should also have access to the personal files like those in the same collaborative work team.

XACML offers extensibility and pluggability which enables the policy presented in this work to be not only a standalone policy, but it could also be a small part of a larger collection of policies. Possible extensions of the base collaboration policy could, for example, by sub-roles (Figure 4) of each primary collaboration roles. This could give even more granularity for specific cases for example if a medical employee in the *management* team role.

B. Conclusions and Future work

The WBAC model was proposed by introducing the team role concept and modifying the user role assignment model from RBAC and ABAC works. The team role of each team member will subsequently determine the extent of access given. Moreover, the level of fine-grained control of access (granularity) to objects that can be authorized to healthcare providers is managed and controlled based on the job required.

The WBAC model utilizes role, team role and WBAC policies to perform an access control evaluation process. First, it checks the access request to verify whether the requesting user possesses a valid role specified in the system. If the requesting user holds the right role, WBAC will check the permission associated with the role and then inspect the rule(s) within the main WBAC policies for additional constraints on access. In other models such as RBAC, failure in this stage results in the complete termination of the decision process. WBAC, however, treats this differently. If the requesting user does not hold a valid role (in most cases, the requesting user might be an outsider who is invited to collaborative work and does not hold a role in the organization), WBAC investigates further to determine whether the requesting user is part of the collaborative work. If so, the respective user's team role is extracted and examined for whether the requesting user possesses a valid team role over the resource. WBAC also checks the permission associated with the team role and checks the rule(s) within WBAC collaborative policies for additional constraints on access.

In the future, the plan is to develop and prototype the WBAC functionality to understand the possible difficulties in managing the model during actual implementation; model performance validity could also be evaluated in terms of resource consumption, e.g., time and computational capability. In additional, future research is required to incrementally develop additional types of constraints and policies, to further investigate how the WBAC and access delegation can be enriched to support the various needs on information access management in case of emergency (break-glass policy [71–73]), and to examine the generalizability of the enhanced WBAC model for other applications in healthcare environments such as clinical education and biomedical research.

Acknowledgment

The authors would like to thank Geir M. Køien for the support in investigating and typesetting this work.

References

[1] M. Abomhara and H. Nergaard, "Modeling of work-based access control for cooperative healthcare systems with xacml," in Proceedings of the Fifth International Conference on Global Health Challenges (GLOBAL HEALTH 2016), 2016, pp. 14–21.

[2] A. C. Myers and B. Liskov, A decentralized model for information flow control. ACM, 1997, vol. 31, no. 5.

[3] D. Hedin and A. Sabelfeld, "A perspective on information-flow control," in Proceedings of the 2011 Marktoberdorf Summer School, 2011.

[4] M. Abomhara and G. M. Køien, "Cyber security and the internet of things: Vulnerabilities, threats, intruders and attacks," Journal of Cyber Security, vol. 4, pp. 65–88.

[5] P. Samarati and S. C. de Vimercati, "Access control: Policies, models, and mechanisms," in International School on Foundations of Security Analysis and Design, vol. 2171. Springer, 2001, pp. 137–196.

[6] P. Li, Y. Mao, and S. Zdancewic, "Information integrity policies," in In Proceedings of the Workshop on Formal Aspects in Security & Trust (FAST. Citeseer, 2003.

[7] P. K. Sinha, G. Sunder, P. Bendale, M. Mantri, and A. Dande, Electronic health record: standards, coding systems, frameworks, and infrastructures. John Wiley & Sons, 2012.

[8] J. H. Carter, Electronic health records: a guide for clinicians and administrators. American College of Physicians (ACP) Press, 2008.

[9] R. S. Sandhu, E. J. Coyne, H. L. Feinstein, and C. E. Youman, "Role-based access control models," Computer, no. 2, 1996, pp. 38–47.

[10] D. F. Ferraiolo, R. Sandhu, S. Gavrila, D. R. Kuhn, and R. Chandramouli, "Proposed nist standard for role-based access control," ACM Transactions on Information and System Security (TISSEC), vol. 4, no. 3, 2001, pp. 224–274.

[11] D. Ferraiolo, D. R. Kuhn, and R. Chandramouli, Role-based access control. Artech House, 2003.

[12] V. C. Hu, D. Ferraiolo, R. Kuhn, A. Schnitzer, K. Sandlin, R. Miller, and K. Scarfone, "Guide to attribute based access control (abac) definition and considerations," NIST Special Publication, vol. 800, 2014, p. 162.

[13] D. Ferraiolo, R. Chandramouli, V. Hu, and R. Kuhn, "A comparison of attribute based access control (abac) standards for data service applications:extensible access control markup language (xacml) and next generation access control (ngac)," NIST Special Publication (800-178), vol. 800, no. 178, 2016. [Online]. Available: http://nvlpubs.nist.gov/nistpubs/SpecialPublications/NIST.SP.800-178.pdf

[14] M. H. Kang, J. S. Park, and J. N. Froscher, "Access control mechanisms for inter-organizational workflow," in Proceedings of the sixth ACM symposium on Access control models and technologies. ACM, 2001, pp. 66–74.

[15] S. Oh and S. Park, "Task–role-based access control model," Information systems, vol. 28, no. 6, 2003, pp. 533–562.

[16] R. A. Shaikh, K. Adi, L. Logrippo, and S. Mankovski, "Inconsistency detection method for access control policies," in Information Assurance and Security (IAS), 2010 Sixth International Conference on. IEEE, 2010, pp. 204–209.

[17] X. H. Le, T. Doll, M. Barbosu, A. Luque, and D. Wang, "An enhancement of the role-based access control model to facilitate information access management in context of team collaboration and workflow," Journal of biomedical informatics, vol. 45, no. 6, 2012, pp. 1084–1107.

[18] W. Wang, "Team-and-role-based organizational context and access control for cooperative hypermedia environments," in Proceedings of the tenth ACM Conference on Hypertext and hypermedia: returning to our diverse roots: returning to our diverse roots. ACM, 1999, pp. 37–46.

[19] C. E. Rubio-Medrano, C. D'Souza, and G.-J. Ahn, "Supporting secure collaborations with attribute-based access control," in Collaborative Computing: Networking, Applications and Worksharing (Collaboratecom), 2013 9th International Conference Conference on. IEEE, 2013, pp. 525–530.

[20] Q. Li, X. Zhang, M. Xu, and J. Wu, "Towards secure dynamic collaborations with group-based rbac model," computers & security, vol. 28, no. 5, 2009, pp. 260–275.

[21] M. Abomhara, H. Yang, and G. M. Køien, "Access control model for cooperative healthcare environments: Modeling and verification," in 2016 IEEE International Conference on Healthcare Informatics (ICHI). IEEE, 2016, pp. 46–54.

[22] M. Abomhara and H. Yang, "Attribute-based authenticated access for secure sharing of healthcare records in collaborative environments," in Proceedings of the Eighth International Conference on eHealth, Telemedicine, and Social Medicine (eTELEMED 2016), 2016, pp. 138–144.

[23] H. Shen and P. Dewan, "Access control for collaborative environments," in Proceedings of the 1992 ACM conference on Computer-supported cooperative work. ACM, 1992, pp. 51–58.

[24] M. Abomhara and G. M. Køien, "Towards an access control model for collaborative healthcare systems," in Proceedings of the 9th International Joint Conference on Biomedical Engineering Systems and Technologies (BIOSTEC 2016), vol. 5, 2016, pp. 213–222.

[25] M. Abomhara and M. B. Lazrag, "Uml/ocl-based modeling of work-based access control policies for collaborative healthcare systems," in 2016 IEEE 18th International Conference on e-Health Networking, Applications and Services (Healthcom). IEEE, 2016, pp. 1–6.

[26] M. Abomhara, H. Yang, G. M. Køien, and M. B. Lazreg, "Work-based access control model for cooperative healthcare environments: Formal specification and verification," Journal of Healthcare Informatics Research, 2017, pp. 1–33. [Online]. Available: http://dx.doi.org/10.1007/s41666-017-0004-7

[27] R. M. Belbin, Team roles at work. Routledge, 2012.

[28] R. Meredith Belbin, "Management teams: Why they succeed or fail," Human Resource Management International Digest, vol. 19, no. 3, 2010.

[29] T. Moses et al., "Extensible access control markup language (xacml) version 2.0," Oasis Standard, vol. 200502, 2005.

[30] D. Boneh, E. Shen, and B. Waters, "Strongly unforgeable signatures based on computational diffie-hellman," in International Workshop on Public Key Cryptography. Springer, 2006, pp. 229–240.

[31] H. K. Maji, M. Prabhakaran, and M. Rosulek, "Attribute-based signatures," in cryptographers Track at the RSA Conference. Springer, 2011, pp. 376–392.

[32] M. Abomhara and H. Yang, "Collaborative and secure sharing of healthcare records using attribute-based authenticated access," International Journal on Advances in Security Volume 9, Number 3 & 4, 2016, pp. 148–195.

[33] C. Borrill, M. West, D. Shapiro, and A. Rees, "Team working and effectiveness in health care," British Journal of Healthcare Management, vol. 6, no. 8, 2000, pp. 364–371.

[34] C. Taylor, A. J. Munro, R. Glynne-Jones, C. Griffith, P. Trevatt, M. Richards, and A. J. Ramirez, "Multidisciplinary team working in cancer: what is the evidence?" BMJ, vol. 340, 2010, p. c951.

[35] P. Mitchell, M. Wynia, R. Golden, B. McNellis, S. Okun, C. E. Webb, V. Rohrbach, and I. Von Kohorn, "Core principles & values of effective team-based health care," Washington, DC: Institute of Medicine, 2012.

[36] S. J. Dwyer III, A. C. Weaver, and K. K. Hughes, "Health insurance portability and accountability act," Security Issues in the Digital Medical Enterprise, vol. 72, no. 2, 2004, pp. 9–18.

[37] R. Nosowsky and T. J. Giordano, "The health insurance portability and accountability act of 1996 (hipaa) privacy rule: implications for clinical research," Annu. Rev. Med., vol. 57, 2006, pp. 575–590.

[38] R. Zhang and L. Liu, "Security models and requirements for healthcare application clouds," in Cloud Computing (CLOUD), 2010 IEEE 3rd International Conference on. IEEE, 2010, pp. 268–275.

[39] B. Fabian, T. Ermakova, and P. Junghanns, "Collaborative and secure sharing of healthcare data in multi-clouds," Information Systems, vol. 48, 2015, pp. 132–150.

[40] F. T. Alotaiby and J. X. Chen, "A model for team-based access control (tmac 2004)," in Information Technology: Coding and Computing, 2004. Proceedings. ITCC 2004. International Conference on, vol. 1. IEEE, 2004, pp. 450–454.

[41] G. J. Annas, "Hipaa regulations–a new era of medical-record privacy?" The New England journal of medicine, vol. 348, no. 15, Apr 10 2003, pp. 1486–90. [Online]. Available: https://search.proquest.com/docview/223930472?accountid=45259

[42] J. L. Agris, "Extending the minimum necessary standard to uses and disclosures for treatment: Currents in contemporary bioethics," The Journal of Law, Medicine & Ethics, vol. 42, no. 2, 2014, pp. 263–267.

[43] D. Ferraiolo, J. Cugini, and D. R. Kuhn, "Role-based access control (rbac): Features and motivations," in Proceedings of 11th annual computer security application conference, 1995, pp. 241–48.

[44] M. A. Al-Kahtani and R. Sandhu, "A model for attribute-based user-role assignment," in Computer Security Applications Conference, 2002. Proceedings. 18th Annual. IEEE, 2002, pp. 353–362.

[45] S. Alshehri and R. K. Raj, "Secure access control for health information sharing systems," in Healthcare Informatics (ICHI), 2013 IEEE International Conference on. IEEE, 2013, pp. 277–286.

[46] R. K. Thomas, "Team-based access control (tmac): a primitive for applying role-based access controls in collaborative environments," in Proceedings of the second ACM workshop on Role-based access control. ACM, 1997, pp. 13–19.

[47] U. D. of Health, H. Services et al., "Hipaa privacy rule and sharing information related to mental health." [Online]. Available: http://www.hhs.gov/hipaa/for-professionals/special-topics/mental-health/

[48] R. Gajanayake, R. Iannella, and T. Sahama, "Privacy oriented access control for electronic health records," electronic Journal of Health Informatics, vol. 8, no. 2, 2014, p. 15.

[49] S. A. Wheelan, Creating effective teams: A guide for members and leaders. Sage Publications, Inc, 1999.

[50] C. Borrill and M. West, "Developing team working in health care: A guide for managers," Aston: Aston Centre for Health Service Organisation Research, Aston University, 2001.

[51] A. Zambouri, "Preoperative evaluation and preparation for anesthesia and surgery," Hippokratia, vol. 11, no. 1, 2007, pp. 13–21.

[52] K. Asif, S. I. Ahamed, and N. Talukder, "Avoiding privacy violation for resource sharing in ad hoc networks of pervasive computing environment," in Proceedings of the 31st Annual International Computer Software and Applications Conference-Volume 02. IEEE Computer Society, 2007, pp. 269–274.

[53] O. R. Kurkovsky, O. Rivera, and J. Bhalodi, "Classification of privacy management techniques in pervasive computing," International Journal of u-and e-Service, Science and Technology, vol. 11, no. 1, 2007, pp. 55–71.

[54] J. Thomas et al., "Medical records and issues in negligence," Indian Journal of Urology, vol. 25, no. 3, 2009, p. 384.

[55] M. A. Rothstein, "Hipaa privacy rule 2.0," The Journal of Law, Medicine & Ethics, vol. 41, no. 2, 2013, pp. 525–528.

[56] C. for Disease Control, Prevention et al., "Hipaa privacy rule and public health. guidance from cdc and the us department of health and human services," MMWR: Morbidity and mortality weekly report, vol. 52, no. Suppl. 1, 2003, pp. 1–17.

[57] U. D. of Health, H. Services et al., "Summary of the hipaa privacy rule," Washington, DC: Department of Health and Human Services, 2003.

[58] J. F. Alqatawna, E. Rissanen, and B. Sadighi, "Overriding of access control in xacml," in Policies for Distributed Systems and Networks, 2007. POLICY'07. Eighth IEEE International Workshop on. IEEE, 2007, pp. 87–95.

[59] A. X. Liu, F. Chen, J. Hwang, and T. Xie, "Designing fast and scalable xacml policy evaluation engines," Computers, IEEE Transactions on, vol. 60, no. 12, 2011, pp. 1802–1817.

[60] V. C. Hu, D. Ferraiolo, R. Kuhn, A. R. Friedman, A. J. Lang, M. M. Cogdell, A. Schnitzer, K. Sandlin, R. Miller, K. Scarfone et al., "Guide to attribute based access control (abac) definition and considerations (draft)," NIST Special Publication, vol. 800, 2013, p. 162.

[61] N. Li, Q. Wang, W. Qardaji, E. Bertino, P. Rao, J. Lobo, and D. Lin, "Access control policy combining: theory meets practice," in Proceedings of the 14th ACM symposium on Access control models and technologies. ACM, 2009, pp. 135–144.

[62] S. Microsystems, "Official project web site sun's xacml implementation." [Online]. Available: http://www.sunxacml.sourceforge.net

[63] B. Alhaqbani and C. Fidge, "Access control requirements for processing electronic health records," in Business Process Management Workshops. Springer, 2008, pp. 371–382.

[64] W. Tolone, G.-J. Ahn, T. Pai, and S.-P. Hong, "Access control in collaborative systems," ACM Computing Surveys (CSUR), vol. 37, no. 1, 2005, pp. 29–41.

[65] R. K. Thomas and R. S. Sandhu, "Task-based authorization controls (tbac): A family of models for active and enterprise-oriented authorization management," in Database Security XI. Springer, 1998, pp. 166–181.

[66] C. K. Georgiadis, I. Mavridis, G. Pangalos, and R. K. Thomas, "Flexible team-based access control using contexts," in Proceedings of the sixth ACM symposium on Access control models and technologies. ACM, 2001, pp. 21–27.

[67] W. Zhou and C. Meinel, "Team and task based rbac access control model," in Network Operations and Management Symposium, 2007. LANOMS 2007. Latin American. IEEE, 2007, pp. 84–94.

[68] Q. Li, M. Xu, and X. Zhang, "Towards a group-based rbac model and decentralized user-role administration," in Distributed Computing Systems Workshops, 2008. ICDCS'08. 28th International Conference on. IEEE, 2008, pp. 441–446.

[69] T. Gould, D. Crosby, M. Harmer, S. Lloyd, J. Lunn, G. Rees, D. Roberts, and J. Webster, "Policy for controlling pain after surgery: effect of sequential changes in management." Bmj, vol. 305, no. 6863, 1992, pp. 1187–1193.

[70] A. Mohammad, T. Khdour, G. Kanaan, R. Kanaan, and S. Ahmad, "Analysis of existing access control models from web services applications perspective," J. Comput, vol. 3, 2011, pp. 10–16.

[71] A. D. Brucker and H. Petritsch, "Extending access control models with break-glass," in Proceedings of the 14th ACM symposium on Access control models and technologies. ACM, 2009, pp. 197–206.

[72] M. Hafner, M. Memon, and M. Alam, "Modeling and enforcing advanced access control policies in healthcare systems with sectet," in Models in Software Engineering. Springer, 2007, pp. 132–144.

[73] S. Marinovic, R. Craven, J. Ma, and N. Dulay, "Rumpole: a flexible break-glass access control model," in Proceedings of the 16th ACM symposium on Access control models and technologies. ACM, 2011, pp. 73–82.